MIND TOOLS

BOOKS BY RUDY RUCKER

NONFICTION

Geometry, Relativity and the Fourth Dimension

Infinity and the Mind: The Science and Philosophy of the Infinite

The Fourth Dimension: A Guided Tour of the Higher Universes

Mind Tools: The Five Levels of Mathematical Reality

FICTION

Spacetime Donuts

White Light, or, What is Cantor's Continuum Problem?

Software

The Fifty-Seventh Franz Kafka

The Sex Sphere

Master of Space and Time

The Secret of Life

Wetware

Rudy Rucker

MIND TOOLS

The Five Levels of Mathematical Reality

Houghton Mifflin Company • Boston

Library of Congress Cataloging-in-Publication Data

Rucker, Rudy v. B. (Rudy von Bitter), date.
Mind tools.

Includes index.
1. Information theory in mathematics. 2. Mathematics — Philosophy. I. Title.
QA10.4.R83 1987 510′.1 86-27790
ISBN 0-395-38315-3
ISBN 0-395-46810-8 (pbk.)

Printed in the United States of America

P 10 9 8 7 6 5 4 3 2

Unless otherwise credited, all artwork is by the Design Group — Nancy Blackwell, Susan Micklem, and Sarah Micklem. "Ye shalt gain in Kudzu."

The computer-generated fractal surface on page 176 is from *The Fractal Geometry of Nature* by Benoit B. Mandelbrot. Copyright © 1977, 1982, 1983. Used by permission of W. H. Freeman and Company.

For Sylvia, with love

CONTENTS

INTRODUCTION
THE FIVE MODES OF THOUGHT

1 NUMBER

2 SPACE

3 LOGIC

4 INFINITY AND INFORMATION

INTRODUCTION
THE FIVE MODES OF THOUGHT

Mathematics as Information

The world is colors and motion, feelings and thought . . . and what does math have to do with it? Not much, if "math" means being bored in high school, but in truth mathematics is the one universal science. Mathematics is the study of pure pattern, and everything in the cosmos is a kind of pattern.

The patterns of mathematics can be roughly grouped into five archetypes: Number, Space, Logic, Infinity, and Information. *Mind Tools* is primarily about information, the newest of these archetypes. The book consists of this Introduction and a chapter each about information in terms of number, space, logic, and infinity.

Just to give an idea of what we'll be talking about, let me pick a specific object and show how it can be thought of, as a mathematical pattern, in five different ways. Let's use your right hand.

1. Hand as Number. At the most superficial level, a hand is an example of the number 5. Looking at details, you notice that your hand has a certain number of hairs and a certain number of wrinkles. The fingers have specific numerical lengths in millimeters. The area of each of your fingernails can be calculated, as can its mass. Internal measurements on your hand could produce a lot more numbers: temperatures, blood flow rates, electrical conductivity, salinity, etc. Your hand codes up a whole lot of numbers.

2. Hand as Space. Your hand is an object in three-dimensional space. It has no holes in it, and it is connected to your body. The skin's curved, two-dimensional surface is convex in some regions and concave in others. The hand's blood vessels form a branching one-dimensional pattern. The bulge of your thumb muscle is approximately ellipsoidal, and your fingers resemble the frustums of cones. Your fingernails are flattened paraboloids, and your epithelial cells are cylindrical. Your hand is a sample case of space patterns.

3. **Hand as Logic.** Your hand's muscles, bones, and tendons make up a kind of machine, and machines are special sorts of logical patterns. If you pull this tendon here, then that bone over there moves. Aside from mechanics, the hand has various behavior patterns that fit together in a logical way. If your hand touches fire, it jerks back. If it touches a bunny, it pets. If it clenches, its knuckles get white. If it digs in dirt, its nails get black. Your logical knowledge about your hand could fill a hefty Owner's Manual.

4. **Hand as Infinity.** Abstractly speaking, your hand takes up infinitely many mathematical space points. As a practical matter, smaller and smaller size scales reveal more and more structure. Close up, your skin's surface is an endlessly complex pattern of the type known as "fractal." What you know about your hand relates to what you know about a ramifying net of other concepts — it is hard to disentangle your hand from the infinite sea of all knowledge. Another kind of infinitude arises from the fact that your hand is part of *you,* and a person's living essence is closely related to the paradoxical infinities of set theory (the mathematician's version of theology).

5. **Hand as Information.** Your hand is designed according to certain instructions coded up in your DNA. The length of these instructions gives a measure of the amount of information in your hand. During the course of its life, your hand has been subject to various random influences that have left scars, freckles, and so on; we might want to include these influences in our measure of your hand's information. One way to do this would be to tie your hand's information content to the number of questions I have to ask in order to build a replica of it. Still another way of measuring your hand's information is to estimate the length of the shortest computer program that would answer any possible question about your hand.

You can think of your hand as made of numbers, of space patterns, of logical connections, of infinite complexities, or of information bits. Each of these complementary thought modes has its use. In the rest of this Introduction I will explain how and why mathematics has evolved the five modes — number, space, logic, infinity, and information — and I will be talking about how these five modes relate to the five basic psychological activities: perception, emotion, thought, intuition, and communication.

Number and Space

Some things vary in a stepwise fashion — the number of people in a family, the number of sheep in a flock, the number of pebbles in a pouch. These are groups of discrete things about which we can ask, "How many?" Other things vary smoothly — distance, age, weight. Here the basic question is, "How much?"

The first kind of magnitude might be called *spotty* and the second kind called *smooth*. The study of spotty magnitudes leads to numbers and arithmetic, while the study of smooth magnitudes leads to notions of length and geometry. Counting up spots leads to the mathematical realm known as "number"; working with smooth quantities leads to the kingdom called "space." The classic example of these two kinds of patterns is the night sky: Staring upward, we see the stars as spots against the smooth black background. If we focus on the individual stars we are thinking in terms of number, but if we start "connecting the dots" and seeing constellations, then we are thinking in terms of space.

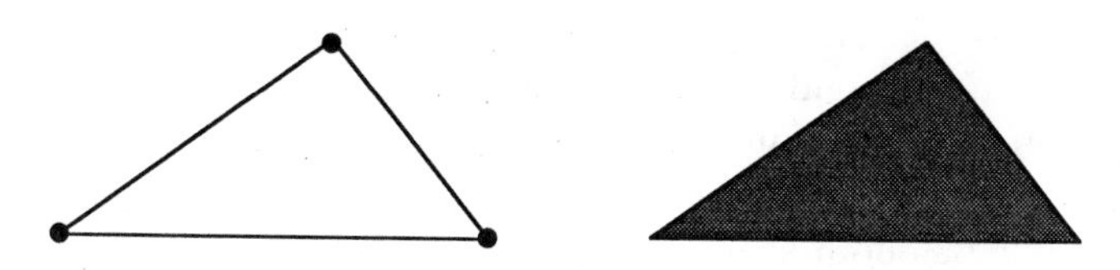

Fig. 1 A triangle as three dots versus a triangle as a piece of space.

The number–space distinction is extremely basic. Together the pair make up what the Greeks called a "dyad," or pair of opposing concepts. In Fig. 2 I have made up a table of some dyads related to the number–space dyad. It is easy to think of other distinctions that seem to fall into the same basic pattern. In human relations you can emphasize either the roles of various individuals, or the importance of the overall society. In Psychology we can talk about either various elementary perceptions or the emotions that link them. A mountainside is covered with trees or by a forest. A piece of music can be

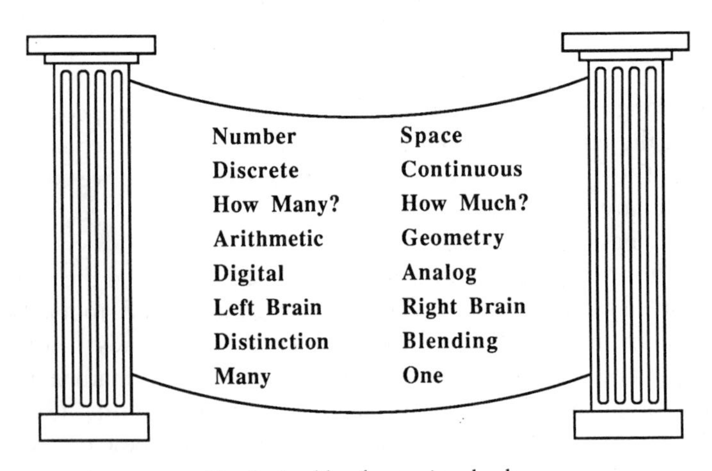

Fig. 2 A table of opposing dyads.

thought of as separate notes, or as a flowing melody. The world can be viewed as a collection of distinct things or as a single organic whole.

Which takes intellectual priority, number or space? Neither. Smooth-seeming matter is said to be made up of atoms, scattered about like little spots, but the chunky little atoms can be thought of as bumps in the smooth fabric of space. Pushing still further, we find some thinkers breaking smooth space into distinct quanta, which are in turn represented as smooth mathematical functions. The smooth underlies the spotty, and the spotty underlies the smooth. Distinct objects are located in the same smooth space, but smooth space is made up of distinct locations. There is no real priority; the two modes of existence are complementary aspects of reality.

The word "complementarity" was first introduced into philosophy by the quantum physicist Niels Bohr. He used this expression to sum up his belief that basic physical reality is both spotty and smooth. An electron, according to Bohr, is in some respects like a particle (like a number) and in some respects like a wave (like space). At the deepest level of physical reality, things are not definitely spotty or definitely smooth. The ambiguity is a result of neither vagueness nor contradiction. The ambiguity is rather a result of our preconceived notions of "particle" and "wave" not being wholly appropriate at very small size scales.

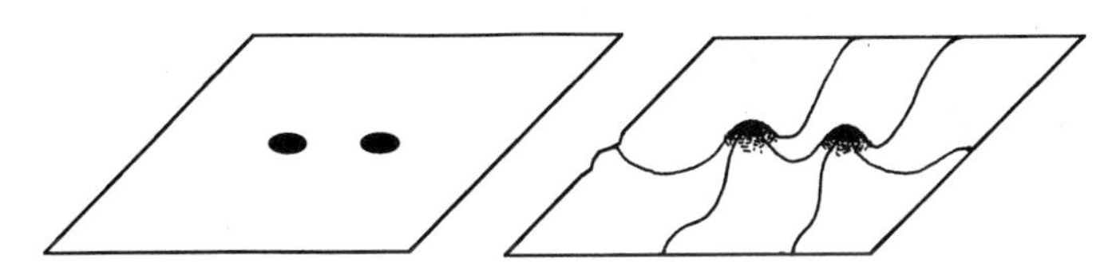

Fig. 3 Particles as lumps versus particles as bumps.

One might also ask whether a person is best thought of as a distinct individual or as a nexus in the web of social interaction. No person exists wholly distinct from human society, so it might seem best to say that the space of society is fundamental. On the other hand, each person can feel like an isolated individual, so maybe the number-like individuals are fundamental. Complementarity says that a person is both individual and social component, and that there is no need to try to separate the two. Reality is one, and language introduces impossible distinctions that need not be made.

Bohr was so committed to the idea of complementarity that he designed himself a coat of arms that includes the yin–yang symbol, in which dark and light areas enfold each other and each contains a part of the other at its core. Bohr's strong belief in complementarity led him to make a singular statement: "A great truth is a statement whose opposite is also a great truth."

Bohr thought of the number–space dyad as being an essential part of reality. Given a dyad, there is always the temptation to believe that if we could only dig a little deeper, we could find a way of explaining one half of the dyad in terms of the other, but the philosophy of complementarity says that there doesn't *have* to be any single fundamental concept. Some aspects of the world are spread out and spotty, like the counting numbers; some aspects of the world are smooth and connected, like space. Complementarity tells us not to try to make the world simpler than it actually is.

It is interesting to realize that two complementary world views seem to be built into our brains. I am referring here to the human brain's allocation of different functions to its two halves.

Fig. 4 Niels Bohr's coat of arms.

A computer is said to be *digital* if it works by manipulating distinct chunks of information. This is opposed to an *analog* computer, which works by the smooth interaction of physical forces. The most familiar examples of analog and digital computers are the two kinds of wrist watches. The old-fashioned analog watch uses a system of gears to move its hands in smooth sweeps, analogous to the flow of time. Loosely speaking, an analog watch is a kind of scale model of the solar system. The newer digital watches count the vibrations of a small crystal, process the count through a series of switches, and display the digits of a number which names the time.

Many of the intellectual tasks a brain performs can be thought of as primarily digital or primarily analog. Arithmetic is certainly a digital activity, and spelling out a printed word is also a basically digital activity. Each of these activities is a "step-at-a-time" process, involving such steps as reading a symbol, finding a meaning for the symbol, and combining two symbols. Singing a song to music, on the other hand, is an analog activity — the speech organs are continuously adjusted to produce tones matching the smoothly varying music. (The digital note patterns of sheet music are but the sluices through which performance flows.) Recognizing a scene in a photograph is also believed to be an analog activity of the brain; the brain seems to see the picture "all at once" rather than to divide it up into lumps of information.

In the 1960s, a variety of experiments involving people with various kinds of brain injuries suggested that, as a rule, the left brain is in charge of digital manipulations and the right brain is in charge of analog activities. Recently this phenomenon has been directly observed by means of PET (Positron Emission Tomography) scans that show that the left brain's metabolism speeds up for digital tasks, while the right brain's activity increases during analog tasks. In other words, the left half of your brain thinks in terms of number, and the right half of your brain thinks in terms of space. (See Fig. 5.)

The actual muscles of the body's right half are controlled by the analytical left brain, while the synthesizing right brain controls the body's left side. Most full-face photographs of people show significant differences between the face's two sides. As a rule, a face's right side will have a more tightly controlled and socially acceptable expression; a face's left side often looks somewhat out of it.

Is it the internal division of brain function that causes us to see the world in terms of the thesis–antithesis pattern of spotty and smooth? Perhaps, but I think the converse is more likely. That is, I think it is more likely that the number–space split is a fundamental feature of reality, and that our brains have evolved so as to be able to deal with both modes of existence.

Logic and Infinity

Mathematics is a universal language, so it is not surprising that in mathematics we find both the spotty and the smooth — more technically known as the discrete and the continuous. A pattern is discrete if it is made up of separate, distinct bits. It is continuous if its parts

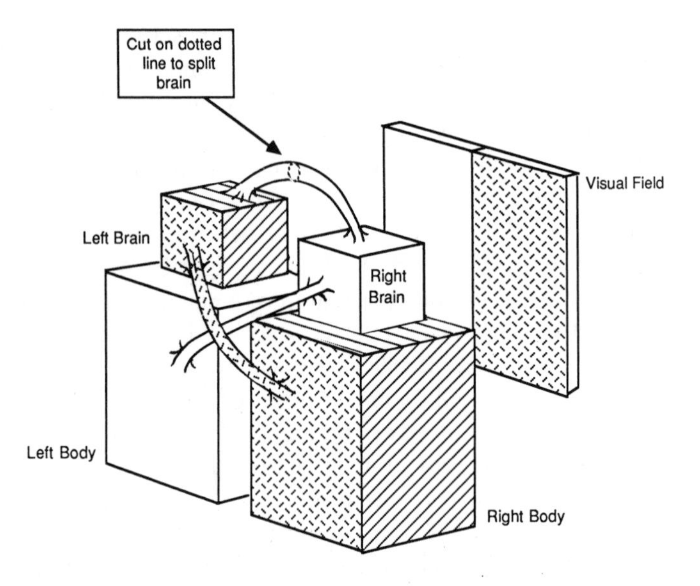

Fig. 5 Our double brain.

blend into an indivisible whole. Viewed as three dots, a triangle is discrete; but viewed as three lines, a triangle is continuous. What gives mathematics so much of its power is that it contains a variety of tools for bridging the gap between space and number. The two most important of these tools are logic and infinity.

"Infinity" is a hypnotic word, suggesting starships, immortality, and endlessness. In older writings it is not unusual for authors to say "The Infinite," where they mean "God." The word "logic" also has a number of colorful associations: cavemen outwitting mastodons; monks analyzing a passage from Aristotle; Ulam and Von Neumann inventing the H-bomb; a robot brokenly asking, "What is Love?"

In reality, logic simply has to do with the idea of letting one general pattern stand for a whole range of special cases. The use of logical techniques enables us to move back and forth between lumpy formulas and smooth mathematical shapes. By a kind of idealization, logic lets a single symbol stand for an ineffably complex reality.

At the lowest level, the logic of mathematics involves using the special symbols for which mathematics is so well known. The equations of algebra are a good example of this kind of low-level mathematical logic. "Algebra" is a wonderful-sounding word for a wonderful skill. Knowing algebra is like knowing some magical language of sorcery — a language in which a few well-chosen words can give one mastery over the snakiest of curves. Like many magical and mathematical words, "algebra" comes from the Arabic, for it was the Arabs

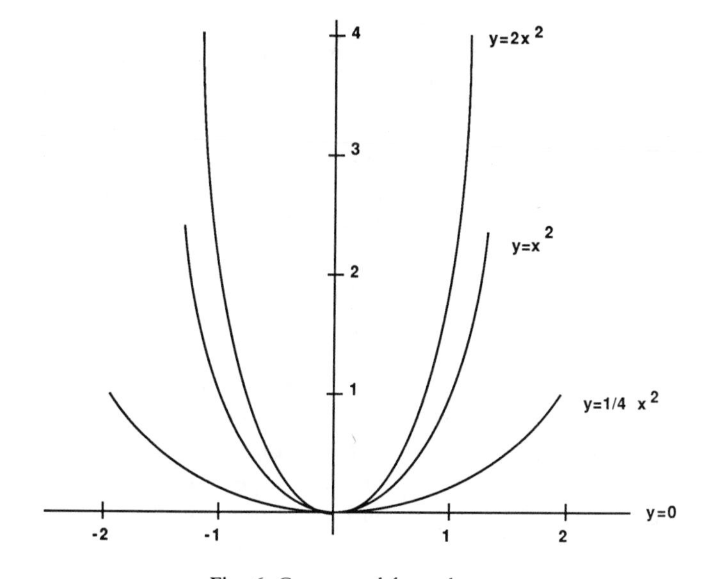

Fig. 6 Curves and formulas.

who kept the Greek mathematical heritage alive during the Dark Ages. "Algebra" comes from the expression *al-jabara,* meaning "binding things together," or "bone-setting." Algebra provides a way for logic to connect the continuous and the discrete. On the one hand, a parabola, say, is a smooth curve in space, yet mathematical reasoning shows us that the parabola can equally well be thought of as a simple algebraic equation — a discrete set of specific symbols.

At a higher level, mathematical logic works through tying individual sentences together into mathematical theories. If we think of a plane filled with points and lines, the initial impression is one of chaos. Once we set down Euclid's five laws (or axioms) about points and lines, however, we have captured a great deal of information about the plane. For purposes of reasoning, the continuous plane is captured by a few rows of discrete symbols.

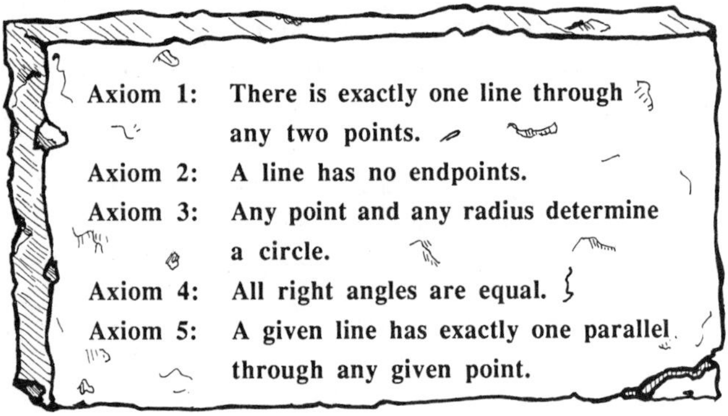

Axiom 1: **There is exactly one line through any two points.**
Axiom 2: **A line has no endpoints.**
Axiom 3: **Any point and any radius determine a circle.**
Axiom 4: **All right angles are equal.**
Axiom 5: **A given line has exactly one parallel through any given point.**

Fig. 7 Euclid's axioms.

Logic synthesizes, but infinity analyzes. Logic combines all the facts about a space pattern into a few symbols; infinity connects number and space by breaking space up into infinitely many distinct points. Like logic, infinity provides a two-way bridge between the continuous and the discrete. We can start with discrete points and put infinitely many of them together to get a continuous space. Moving in the other direction, we can start with some continuous region of space and

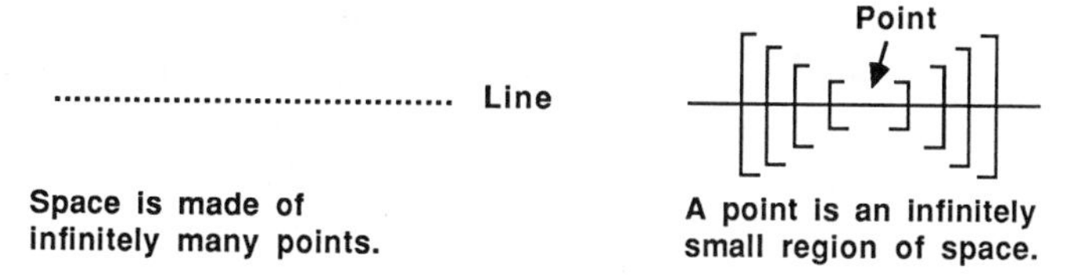

Fig. 8 Infinity is a two-way bridge between the discrete and the continuous.

narrow down to a point by using an infinite nested sequence of approximations.

These two concepts are incorporated into the construction known as the *real number line*. Although I have been talking a lot about the difference between discrete and continuous magnitudes, it is hard for a modern person to realize how really different the two basic kinds of magnitudes are. This is because, very early in our education, we are all taught to identify points on a line with the so-called real numbers. At some time during our high school education, we are taught to measure continuous magnitudes in decimals — to say things like, "My height in meters is 1.5463 . . . ," where, ideally, the ". . ." stands for an endless sequence of more and more precise measurements. The real number system is a concrete example of infinity being used to convert a line's space into decimal numbers.

Calculus uses infinity constantly. Indeed, calculus is sometimes known as Infinitesimal Analysis, where "infinitesimal" means "infinitely small." In calculus we learn to think of a smooth curve as being like a staircase with infinitely many tiny, discrete steps. Thinking of a curve this way makes it possible to define its steepness. This process is known as "differentiation." Another use of infinity peculiar to calculus is the process known as "integration": Given an irregular region whose area we wish to know, calculus finds the area by cutting the region into infinitely many infinitesimal rectangles. (See Fig. 9.)

Although logic and infinity serve as bridges between the discrete and the continuous, looked at on a higher level, they also reflect the gap. Thinking logically is basically a digital, left-brain activity, while talking about infinity is an analog, right-brain process. The two modes of thought are complementary. Mathematicians often use discrete, logical axiom systems to describe various kinds of infinite structures. Logic can never fully encompass the riches of infinity, however. Kurt

Godel proved this in 1930, when he showed that no finite logical system can prove *all* of the true facts about the infinite set of natural numbers.

Psychological Roots of Mathematical Concepts

People often wonder why it is that mathematics is so effective in the sciences. Unlike chess or astrology, mathematics has the curious property of being an intellectual game that really *matters.* Mathematics helps people build computers and cars, TVs and skyscrapers. Mathematics helps predict when the sun will come up and what the weather will be tomorrow. Mathematics has sent people to the moon and back. Why does math work so well?

As I mentioned above, mathematics is a language whose form is universal. There is no such thing as Chinese mathematics or American mathematics; mathematics is the same for everyone. Mathematics consists of concepts imposed on us from without. The ideas of mathematics reflect certain facts about the world as human beings experience it. Just as our bodies have evolved in response to objective

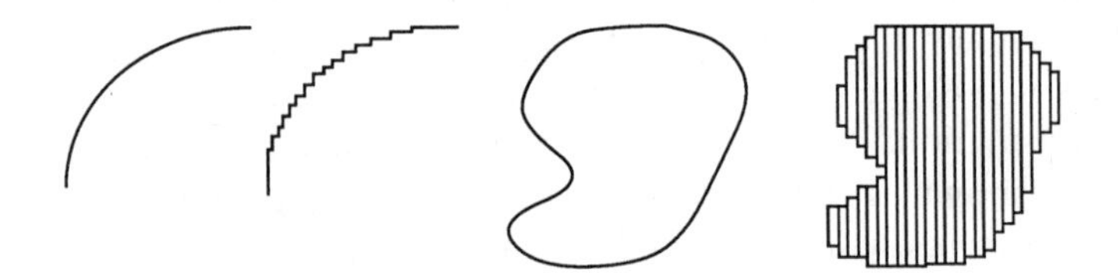

Fig. 9 Differentiation and integration.

conditions imposed by the environment, our ideas have evolved in response to certain other fundamental features of reality.

That distinctions among objects can be made leads to our perception of discreteness. Discreteness leads, in turn, to number. Things do come in lumps, and it is natural to count them.

That smooth transitions can be made leads to our perception of continuity. Things blend into each other, and it is natural to think of them as being in a space that we can measure.

That different kinds of things can resemble each other leads to our perception of similarity. Discussing similarity patterns leads to logic. Various kinds of forms recur, and it is natural to reason about them.

That the world has no obvious boundaries leads to our perception of endlessness. Endlessness leads to infinity. Reality seems inexhaustible, and it is natural to intuit this.

It is worth noting here that the four areas of mathematics — number, space, logic, and infinity — are all treated in most high schools. When students are drilled in arithmetic, they are learning how to manipulate number. Geometry is quite obviously a study of space. Algebra, as was mentioned above, is a type of logic, and calculus, which is often introduced in the twelfth grade, is really a study of infinity.

Number, space, logic, and infinity — the most basic concepts of mathematics. Why are they so fundamental? Because they reflect es-

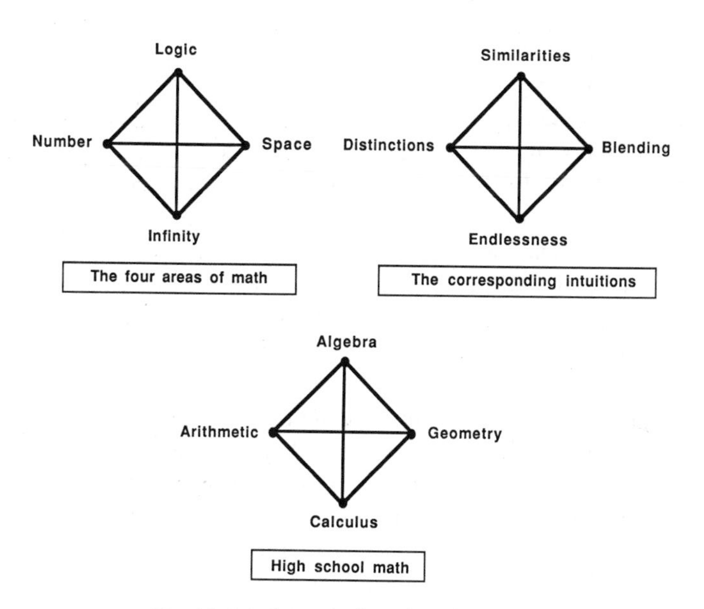

Fig. 10 Mathematical modes of thought.

sential features of our minds and the world around us. Mathematics has evolved from certain simple and universal properties of the world and the human brain. That our mathematics is effective for manipulating concepts is perhaps no more surprising than that our legs are good at walking.

Moving outward, we can use mathematics to change the world around us. Moving inward, we can use mathematics as a guidebook to our own psyches. The early Greeks frequently organized their thoughts in terms of dyads — again, a pair of opposing concepts. The Pythagoreans (who made up one of the earliest schools of mathematics) often drew up "Tables of Opposites" listing related dyads.

Although it's not quite relevant, I can't resist observing here that a DNA molecule is something like a very long table of opposites. The molecule consists of two intertwined "backbones" held together by a long sequence of dyads — matching "base pairs." The DNA molecule reproduces itself by unzipping down the middle, each unzipped half serving as a template for assembling the missing half. In the same way, given half of a table of opposites, it is not too hard to reconstruct the missing half. Does the similarity between a DNA molecule and a table of opposites have any real significance? Yes. The pattern that DNA and the table of opposites share is pervasive enough to be regarded as an archetype, or an important pattern. Even more pervasive is the archetype of the dyad (also known as the number 2).

A dyad is a basically static grouping of concepts — a sort of frozen tug of war. One of G. W. F. Hegel's (who was my great great great grandfather) contributions to philosophy was the idea of grouping concepts into *triads,* which consist of three concepts arranged in the well-known thesis–antithesis–synthesis pattern. The triad is an essentially dynamic grouping, for each synthesis can become the thesis for a new antithesis. The concepts we will be discussing in this book can be grouped into a series of triads, as shown in Fig. 12 on page 18.

Of course this kind of grouping can be forced too hard and should not necessarily be taken very seriously. Any method of organizing our concepts can easily turn into a "Procrustean bed." For those who don't know the story of Procrustes's bed, let me recall that Procrustes was a legendary outlaw who lived in the wilderness near ancient Athens. His house was near a road, and he would invite weary travelers to spend the night in a special bed that he kept for visitors. The catch was this: If you were too tall for the bed, Procrustes would

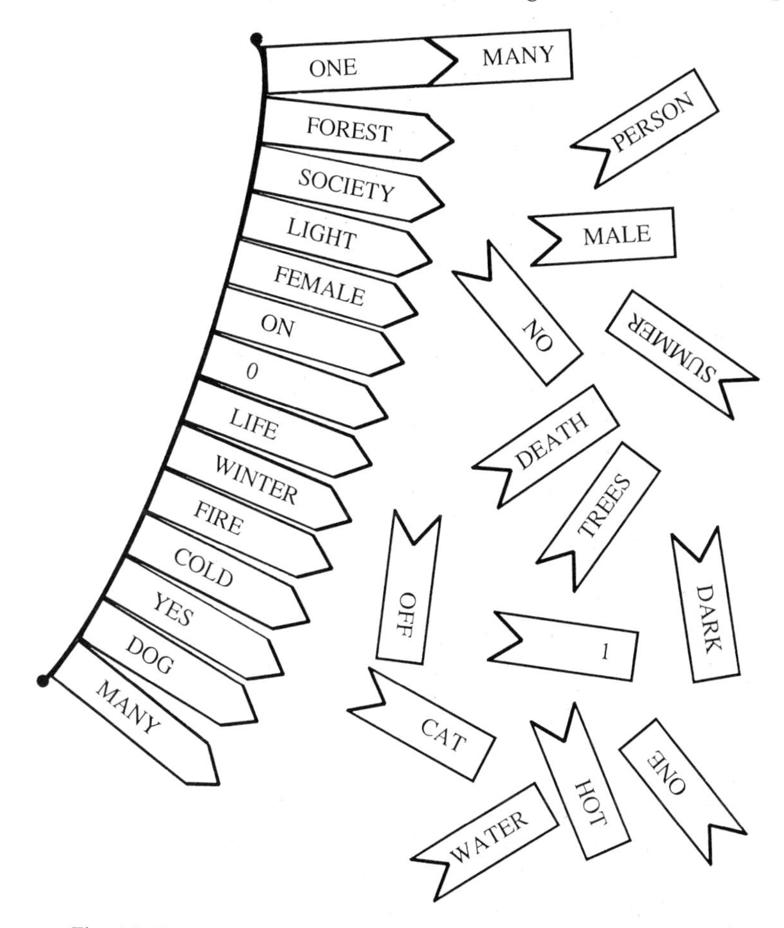

Fig. 11 Can you arrange the loose half-dyads in the right order?

chop off those parts of you that stuck out; and if you were too short for the bed, Procrustes would stretch you by tying your feet to a stake, bending down a tall sapling, and tying its top to your neck!

Just as Hegel goes a step beyond the Greeks, the psychologist C. G. Jung goes a step beyond Hegel. For Jung, the fundamental pattern of thought is not the triad, but the *tetrad,* a balanced, mandala-like arrangement of four concepts, also known as the *quaternity:*

> The quaternity is an archetype of almost universal occurrence. It forms the logical basis for any whole judgment. If one wishes to pass such

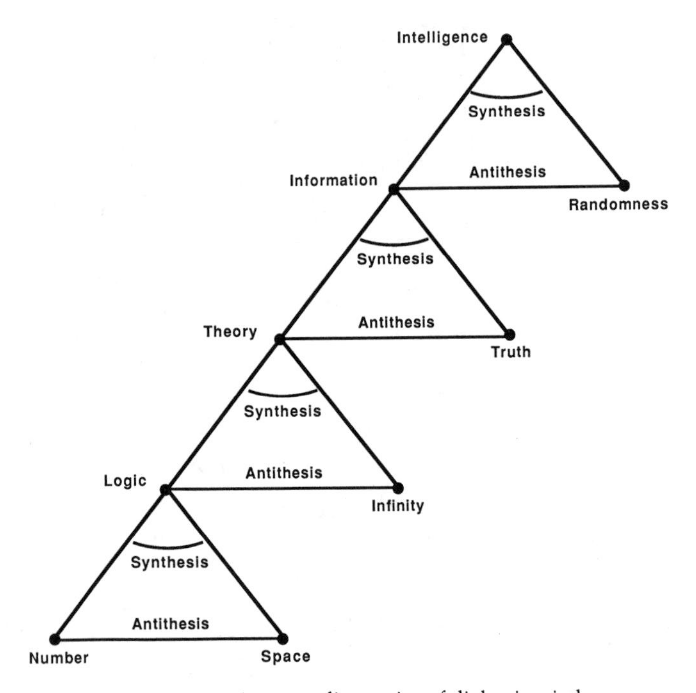

Fig. 12 An ascending series of dialectic triads.

> judgment, it must have this fourfold aspect. For instance, if you want to describe the horizon as a whole, you name the four quarters of heaven. . . . There are always four elements, four prime qualities, four colors, four castes, four ways of spiritual development, etc. So, too, there are four aspects of psychological orientation. . . . In order to orient ourselves, we must have a function which ascertains that something is there (sensation); a second function which establishes *what* it is (thinking); a third function which states whether it suits us or not, whether we wish to accept it or not (feeling); and a fourth function which indicates where it came from and where it is going (intuition). When this has been done, there is nothing more to say. . . . The ideal of completeness is the circle or sphere, but its natural minimal division is a quaternity. (C. G. Jung, *A Psychological Approach to the Dogma of the Trinity,* 1942. In *Collected Works,* Vol. 11, p. 167.)

In Figure 10 I already drew some tetrads (I prefer "tetrad," which Jung also uses sometimes, to the unwieldy "quaternity"); here, in

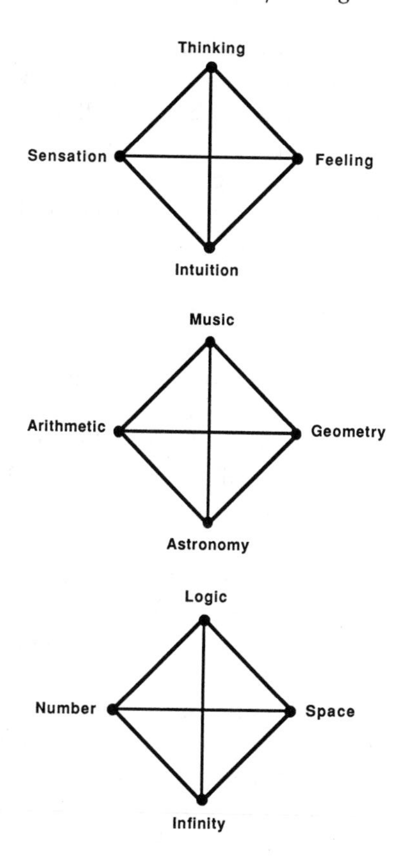

Fig. 13 Jung's psychological tetrad, Plato's "quadrivium," and the math tetrad.

Figure 13 I show our basic tetrad — the tetrad Jung mentions — as well as further tetrads that will be discussed below. Do the parts of Jung's tetrad match the members of our math tetrad in a term-for-term way? As an admirer of the noble Procrustes, I am tempted to say "Yes" and to argue that:

1. *Sensation* ≈ *Number*. What sensation is really about is *making distinctions*. The world of sensations is granular in structure. A red patch is not a green patch. It is this making of distinctions that leads to the world of number.

2. *Feeling* ≈ *Space.* Having a feeling about something involves being connected to it. We stand back as impartial observers to gather our sensations, but to get a feeling for things, we flow out and merge into them. Only in continuous space is such merging possible.

3. *Thinking* ≈ *Logic.* Thinking involves seeing the abstract structures that link our sensations and our feelings. In the process of thinking we look for underlying patterns and compare them. This leads to logic.

4. *Intuition* ≈ *Infinity.* Intuition means getting some deep sense of reality as a whole. Viewing the world as a single, mysterious whole leads quite readily to the formation of the concept of infinity.

Just how long *was* Procrustes's bed, anyway?

Seriously, it is a little surprising how well the two tetrads seem to fit. Until now we have been drawing tetrads as the corners of a square; one might also think of a tetrad as comprising the corners of a three-dimensional *tetrahedron.*

For some reason, tetrahedra aren't as well known as their klutzy Egyptian cousins, the pyramids. The difference between a pyramid and a tetrahedron is that a pyramid is square on the bottom, instead of triangular. What makes the tetrahedron so nice is that no matter which way you turn it, it looks the same. Each corner of a tetrahedron is the same distance from all the other corners. If we think of a triad as an equilateral triangle, then we might say that a tetrad, viewed as a tetrahedron, is made up of four triads.

It would be misleading to say that Hegel and Jung *invented* triads and quaternities. They simply brought these forms to public attention, in much the same way that a chemist might point out that the members of a large group of molecules share a certain kind of atomic architecture. Dyads, triads, and tetrads are *archetypes,* where "archetype" means "a recurrent form of human thought." Triads and tetrads have been around for as long as people have.

Plato, for instance, frequently reasoned dialectically, fitting theses, antitheses, and syntheses together into chains of triads. The Pythagoreans had their own mathematical quaternity: arithmetic, geometry, music, and astronomy.

In the Middle Ages these four branches of mathematics were called the *quadrivium,* and they made up the core of any scientific course of study. Music bridges arithmetic and geometry, in that music consists of discrete notes arranged in continuous time. Astronomy also lies between arithmetic and geometry, in that astronomy concerns discrete bodies moving about in continuous space. Music and astronomy are in some measure opposed to each other, in that songs are created by men's logical minds, while the stars are simply given by the infinite cosmos.

Thought patterns . . . archetypes. "Archetype," by the way, is another of the wonderful words that Jung made up. Let me quote him once again:

> Again and again I encounter the mistaken notion that an archetype is determined in regard to its content, in other words that it is a kind of unconscious idea (if such an expression be admissible). It is necessary to point out once more that archetypes are not determined as regards their content, but only as regards their form and then only to a very limited degree. A primordial image is determined as to its content only when it has become conscious and is therefore filled out with the material of conscious experience. Its form, however, . . . might perhaps be compared to the axial system of a crystal, which, as it were, preforms the crystalline structure in the mother liquid, although it has no material existence of its own. This first appears according to the specific way in which the ions and molecules aggregate. The archetype in itself is empty and purely formal, nothing but a . . . possibility of representation which is given *a priori*. The representations themselves are not inherited, only the forms, and in that respect they correspond in every way to the instincts, which are also determined in form only. The existence of the instincts can no more be proved than the existence of the archetypes, so long as they do not manifest themselves concretely. (C. G. Jung, *The Archetypes and the Collective Unconscious,* 1934.)

It is evident that the thought forms dyad, triad, and tetrad are objectively given archetypes. They correspond to the very basic numbers 2, 3, and 4. The number 5 is also quite basic, and we might suppose there to be a thought form consisting of five related concepts. Let's call this form a pentad. One way of drawing a pentad is as a "quincunx," a legitimate dictionary word meaning "an arrangement of five things with one at each corner and one in the middle of a square."

This is, of course, a flattened picture. Just as the quaternity takes its truest form if we let it pop up into a three-dimensional tetrahedron, it turns out that a pentad takes its most natural shape if we let it sproing out into a four-dimensional "pentahedroid."

I call these shapes "the most natural," because each shape has the

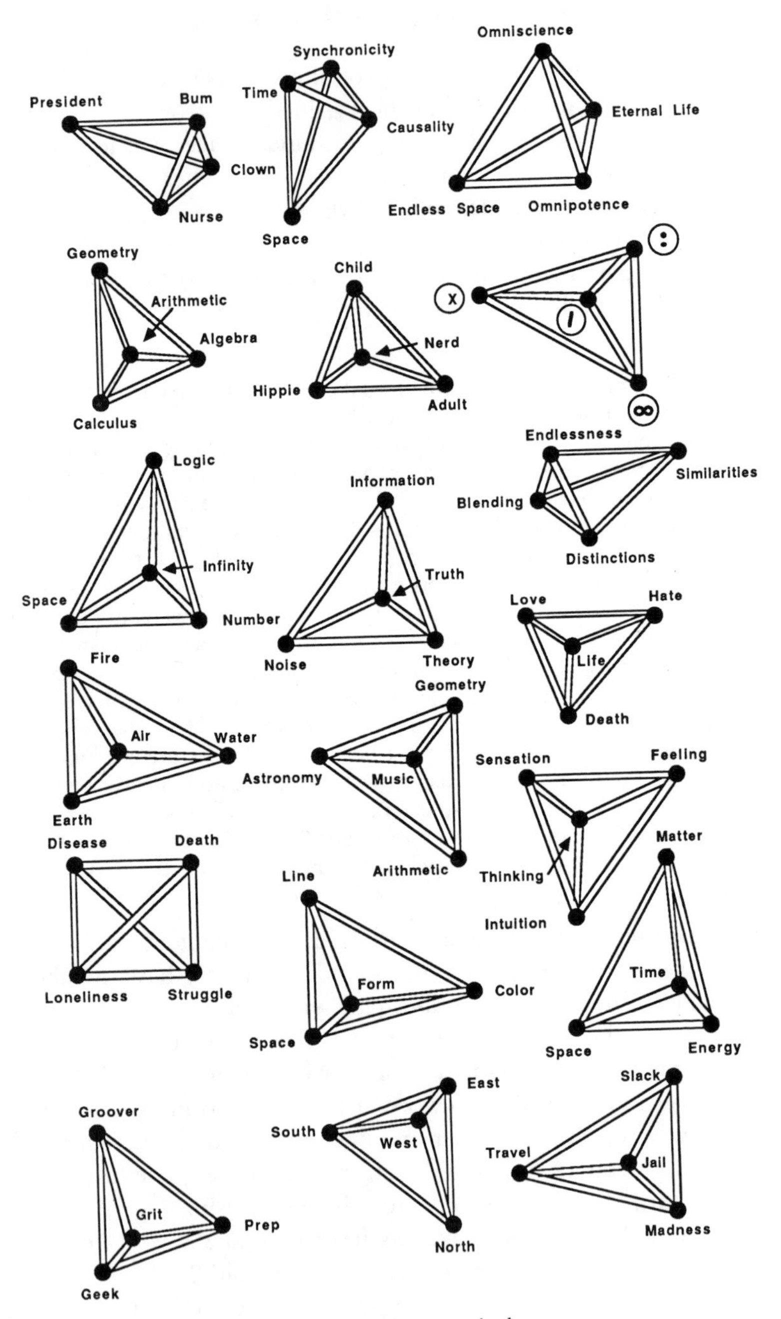

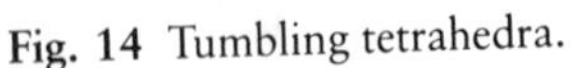
Fig. 14 Tumbling tetrahedra.

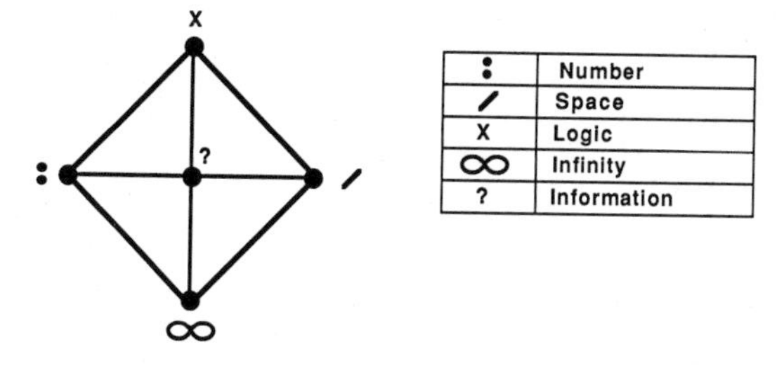

:	Number
/	Space
x	Logic
∞	Infinity
?	Information

Fig. 15 A quincunx.

property that all of its corner points are at the same distance from all the other corner points. Shapes like these are generically called *simplexes.* I should point out that for the picture of the pentahedroid in Figure 16 to be correct, you have to imagine that the central point is pushed away into the fourth dimension a bit, so as to make all the lines equal in length. The pentahedroid is sometimes also called a "five-cell," because it can be thought of as being made up of five equal tetrahedra, just as a tetrahedron can be thought of as being made up of four equal triangles.

Pentahedroids aren't often seen in these parts, but as abstract patterns they do certainly exist, and if an archetype isn't an abstract pattern, then what is?

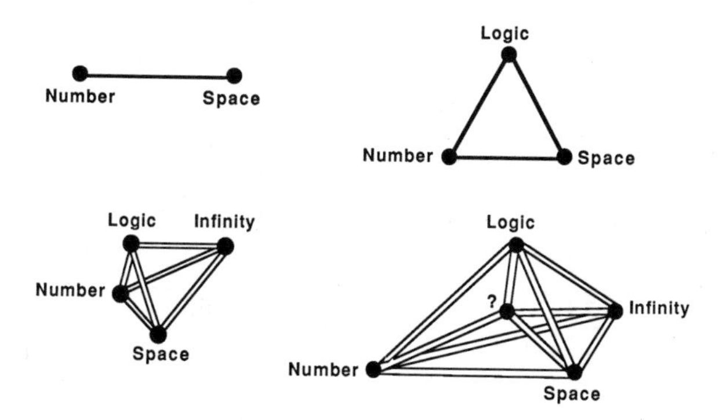

Fig. 16 Simplexes: segment, triangle, tetrahedron, pentahedron.

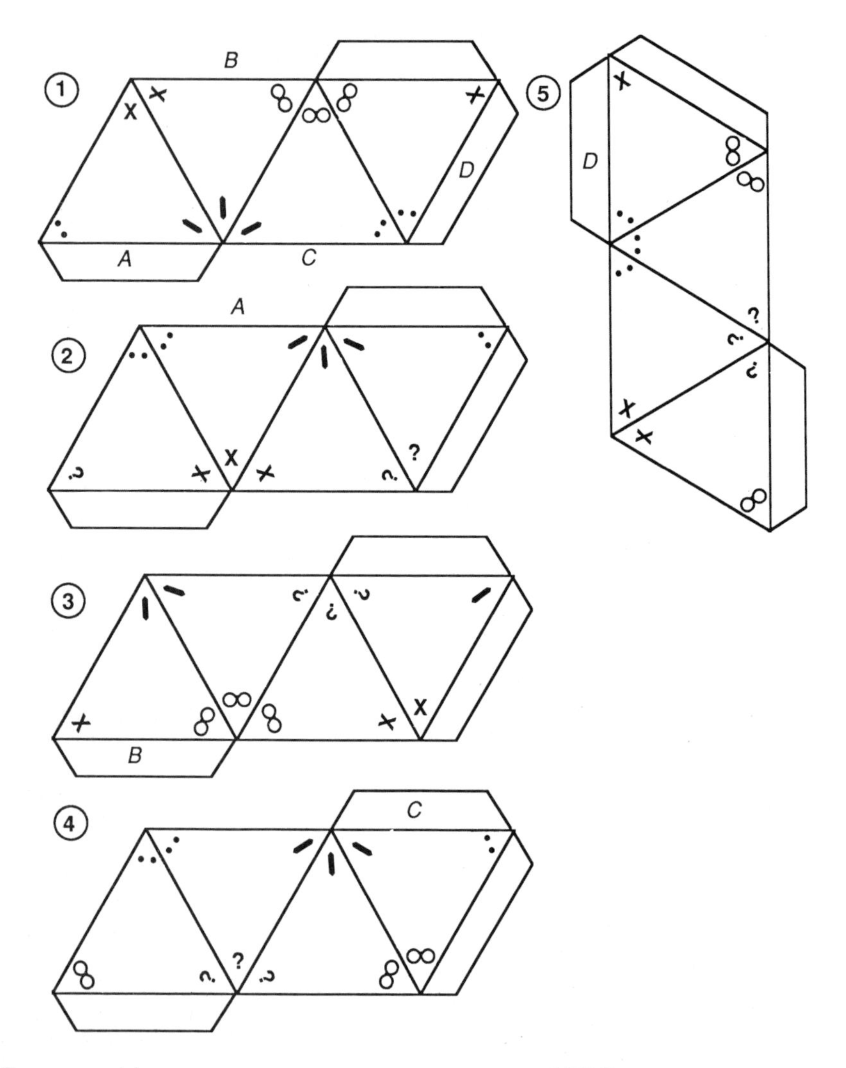

To assemble:
Cut out the 5 patterns and, using rubber cement on glue-flaps, assemble into 5 individual TETRAHEDRA (triangle-based pyramids). Attach each of ② - ⑤ to ① via a scotch-tape = hinged edge so same lettered faces match.

CODE:

:	Number
/	Space
X	Logic
∞	Infinity
?	Information

Fig. 17 The pentahedroid (or "five-cell") of mathematical knowledge.

In any case, what we're looking for is a fifth concept that fits in at equal distances from all four of the concepts we already have. I want to select a fifth mathematical concept as basic as number, space, logic, and infinity. I would also like there to be a corresponding fifth psychological concept, a concept as basic as sensation, feeling, thinking, and intuition. How shall I find these concepts?

It would perhaps not be too much of a distortion to say that the successive eras of human intellectual history in the last thousand years have corresponded to number (the Middle Ages), space (the Renaissance), logic (the Industrial Revolution), and infinity (Modern Times). With the advent of computers, the age of infinity is now drawing to a close. We are entering a new age of mathematics. What is the fifth mathematical concept that best characterizes our new era?

Information.

Information and Communication

Logic connects number and space by reasoning about the similar forms that occur in both areas. Infinity connects number and space by endless processes of approximation. In what way does the concept of information connect number, space, logic, and infinity?

Looked at quite concretely, mathematics is about solving problems. Mathematics turns shapes into areas, conjectures into theorems, equations into solutions. Mathematics turns questions into answers; this is a process of generating information.

Like logic, information is a somewhat ill-defined concept. We can discuss information at three primary levels: message length, complexity, and understanding.

At the simplest level, we feel that information has to do with sending and receiving messages. It seems natural to suppose that the information in a message has to do with the length of the message. A long phone call has more information than a short one; a fat book more information than a thin one.

At the next level, we relate information to the idea of complexity.

Some patterns strike us as simple, not having much information. Other, more structured patterns seem to be telling us more, to carry a lot of information. Five minutes of silence over the phone tells you a lot less than thirty seconds of animated conversation. If a blank page seems uninteresting, it is because it carries little complexity, little information. One might consider relating an object's information content to the length of the shortest list of instructions for building a copy of the object in question. Under this definition, a random-seeming mess would have a lot of "information," as it would be hard to give a short description of it. An unmown, weedy lawn would have more information than a golf green.

At a third level, we think of information as having to do with knowledge and with understanding. This notion of information connects with the idea of messages in a natural way, because it is reasonable to say that knowledge always arises through a dialogue — between two individuals, between a person and the world, or even between two parts of a person's mind. For no matter what branch of knowledge you wish to study, your course of study will ordinarily involve looking at (or listening to, or smelling, or tasting, or feeling) various kinds of objects. Some of the objects you look at will be natural — clouds, waves, trees, and dogs. Your observations will always affect the objects, to a greater or lesser extent, and your observations will leave traces on your own brain. Some of the objects you look at will be people, and some of these people will talk to you and make signs with their hands. You may ask these people questions, and they may answer. Some of the objects you look at will be books about things you are interested in. To digest a book, you will think about it; you will write in its margins; you will argue about it with yourself, or with your mental image of the author. Learning things involves absorbing information — raw information from the physical world, as well as information from other people. Learning things also involves generating information — asking questions, forming memory traces, and writing things down. Whenever you are learning, you are sending and receiving transmissions, generating and absorbing information. You are *communicating.*

Even more than our first four basic concepts, the concept of information currently resists any really precise definition. Relative to information we are in a condition something like the condition of seventeenth-century scientists regarding energy. We know there is an important concept here, a concept with many manifestations, but we do not yet know how to talk about it in exactly the right way.

Despite all this, there is a branch of mathematical science called *information theory*. Appropriately enough, the modern mathematical theory of information was founded by Claude Shannon, an engineer at Bell Laboratories, home of the telephone. Shannon's most important papers, written in the 1940s, arose from work having to do with improving the reliability of long-distance telephone and telegraph lines. Indeed, Shannon originally called his work communication theory, rather than information theory.

How exactly does one go about *measuring* information so as to formulate a scientific theory about it? For Shannon, information is measured in "bits." If I ask you a yes-or-no question, then your answer supplies me with one bit of information. If I play a game of Twenty Questions with you, then, in the process of my asking you twenty yes-or-no questions, I am getting twenty bits of information out of you. A halftone picture in a newspaper can be thought of as having as many bits as it has black-or-white dots. A message in Morse code might be said to have as many bits as there are long-or-short beeps. A flip-flop on-off switch in a computer's memory bank stores up one bit of information. But what about things that don't naturally fall into patterns of yes-or-no choices? How do we assign a bit count to a painting, or to a page of English prose? The Twenty Questions idea is useful here.

Recall that in Twenty Questions the objective is for the asker to figure out which famous person the answerer is thinking of. The asker begins with questions like: "Is the person male?" "Is the person American?" "Is the person presently living?" "Is the person in show business?" It is a little surprising how often the asker can actually figure out which famous person the answerer is thinking of on the basis of only twenty bits of information! (It turns out that Twenty Questions usually works because there are only about one million famous people to choose from. Mathematically inclined readers will recognize that this follows from the fact that one million is approximately equal to two to the twentieth power.)

In general, given any message, the information content of the message is said to be equal to the number of yes-and-no questions one would have to ask to guess the actual message. Obviously, a long written message in English is going to carry a pretty large number of bits of information. Shannon estimated that normal written English has an information content of about 9.14 bits per word, though an English passage that uses a lot of unusual words will have a greater information content. For Shannon, the information content of a message is, in a sense, related to the degree of *unpredictability* of the message.

	No	*Yes*
Does the person's last name start with a letter between A and M?	X	
Does the person's last name start with a letter between N and S?		X
Does the person's last name start with N, O, or P?		X
Does the person's last name start with N or O?	X	
Is the person's age under 30?	X	
Is the person's age under 60?		X
Is the person male?		X
OK. After seven questions, I know that the person is an older man whose last name starts with P. Is he American?		X
Have I ever met him?	X	
Is he a politician?	X	
Is he a performer?	X	
Is he an artist of some sort?		X
Is he a painter?	X	
Is he a novelist?		X
OK. After fourteen questions, I know that the person is an older American male novelist whose last name starts with P. Has he published more than five novels?	X	
Have his books been reviewed in the *New York Times Book Review?*		X
Is it Charles Portis?	X	
Is it Thomas Pynchon?		X

Fig. 18 Thomas Pynchon in eighteen bits of information.

An example: A novel by Vladimir Nabokov carries more information than does a Gothic romance; that is, if a single word is missing from a Nabokov book, it might take a dozen guesses to ascertain the word, whereas a word missing from a Gothic romance can probably be guessed in two tries. This observation is not necessarily a value judgment; it is an objective fact about the kinds of language used in the two types of books. Nabokov uses arcane words; authors of Gothics are expected to avoid such words.

There is, of course, no real need for a message to be made up of words. A message might just as well be a picture, an equation, a wave form, or even a thought. The medium for transmitting the information is known as the *channel*. When we speak to one another, the channel we are using is sound waves. Looked at a bit differently, we speak by varying the air pressure around us. If people had never heard of speech, how strange it would seem! Imagine, by comparison, a race of beings who communicate among themselves by varying the magnetic-field strengths in their vicinity.

The thoughtful reader will see a number of problems with Shannon's Twenty Questions measure of information, and these problems are real. What if the asker is a recluse who has never heard of any of the famous people? What if the asker is Japanese, and you must first teach her English? What if the asker is a giant slug from Saturn accustomed to talking via methane mucus densities? The problem is that one receiver's information is another receiver's random noise.

One of the confusing things about information is that it is, to some extent, a *relative* concept rather than an *absolute* one. The situation is a bit like logic — if I say that something is logically provable, then you can naturally ask, "Provable from *what?*" By the same token, if I say that some pattern contains a lot of information, you may well ask, "Information for *whom?*" At this time, "information" has no universally applicable definition, but the only way we will get anywhere is by trying. More than anything else, *Mind Tools* is an investigation into the nature of information.

Let me say a little more about the question of how information can be said to unify the four traditional branches of mathematics. Looked at in a certain way, most mathematical techniques have to do with rules for finding things out. In grade school we memorize the times tables up to ten times ten, and we learn techniques for working out the values of larger products. In algebra we learn how to solve simple equations in one or two unknowns. When we study Euclidean geometry, we learn how to put axioms together into proofs of theorems. Calculus provides a set of techniques for computing irregular areas and volumes, and so on. Mathematics is, to a large degree, a body of techniques for transforming one kind of information into another. Working a math problem is akin to decoding a message.

The decoding techniques of mathematics are collectively known as *algorithms,* a word derived from the name of the ninth-century Arab mathematician al-Khuwarizmi. Al-Khuwarizmi wrote a book on the art of calculating with the then-unfamiliar "Arabic" number notation that we all use today. In modern times, the science of programming computers is sometimes known as "algorithmics."

Harking back to the general considerations of the last section, what aspects of human psychology might we best relate to information? One answer is communication. Insofar as we do not exist as isolated individuals, we try to communicate with each other. This is a basic psychological activity. Closely related to communication is memory. Insofar as we do not exist solely at one isolated instant in time, we try to remember what happened to us in the past. Communication involves being able to reproduce some information pattern at a distant place; analogously, memory involves being able to reproduce some information pattern at a distant time.

Anyone at the receiving end of a communication is involved with "information processing." I may communicate with you by writing this book, but you need to organize the book's information in terms of mental categories that are significant for you. You must process the information.

Spoken and written communication are, if you stop to think about it, fully as remarkable as telepathy would be. How is it that you can know my thoughts at all, or I yours? You have a thought, you make some marks on a piece of paper, you mail the paper to me, I look at it, and by some mysterious communication algorithm I construct in my own brain a pattern that has the same feel as your original thought.

Information!

A History of Ideas

Nowadays, computers are no more controversial than television. Like them or not, they're a fact of life. The revolution is over. Computers are in our factories, on our desks, and just about everywhere else. Microcomputers are built into most new appliances and toys. What were we so frightened of?

I think the real issue was that the computer revolution forced people to begin viewing the world in a new way. The new world view that computers have spread is this: *everything is information.* It is now considered reasonable to say that, at the deepest, most fundamental level, our world is made of information. Under this new world view, what you see is what you get, and what you say is what you are. For postmodern people, reality is a pattern of information, a pattern in fact-space.

Number, space, logic, infinity, and information. I have already discussed how the five basic concepts relate to each other philosophically and psychologically; now I'd like to use them as the basis for a brief history of ideas. If this is the Information Age, what were the earlier Ages of Western thought? How do they fit together, and how did they start?

1. Middle Ages. The use of written language made it possible to invent and preserve specific names for things. The written name of a thing was a near-magical symbol for the thing itself.

The characteristic activity of the Middle Ages was labeling. Medieval science lists endless strings of special cases. The objects in medieval paintings were used as symbols for various archetypal vices and virtues. Theologians concerned themselves with finding names for levels of beatitude. The thrust of the era was to turn away from the actual world and to shuffle through the stacks of labels.

The activity of labeling is essentially number-oriented. Numbers themselves are simple, universal labels, and numerologists view them this way. The medieval belief that the entire cosmos could be categorized according to preordained schemes is essentially a belief that

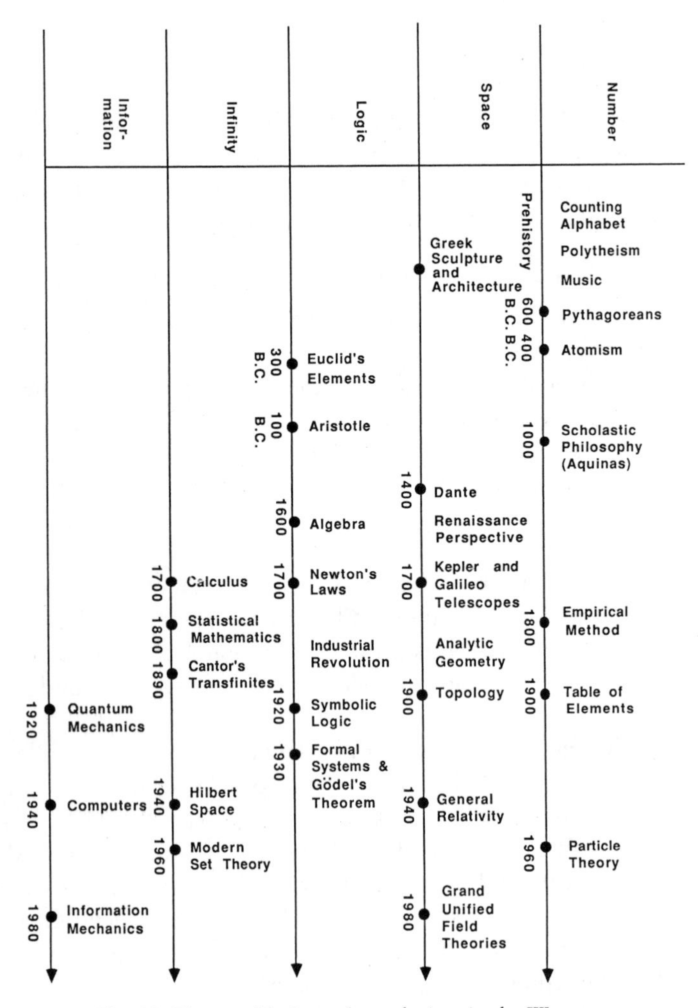

Fig. 19 Five world views through time in the West.

the world is like a heap of numbers. The Middle Ages was the Age of Number.

2. Renaissance. The key fact about Renaissance perspective painting is that physical space is treated in a mathematical way. A medieval painting is a flat screen with icons displayed on it to code up a numberlike message, but a perspective picture ties the picture objects into a single mathematical space. This reflects a radically different way of looking at the world. Instead of being names in God's mind, things became structures in mathematical space. The Renaissance was the Age of Space.

Galileo's astronomical discoveries made it clear that the space between the planets is essentially no different from the space we walk around in. More importantly, his experiments with balls rolling down ramps showed that physical motions are regulated by the same laws in every location. Instead of viewing objects as "acting according to their nature," people began to think of objects as acting in accordance with forces that spread out through space. Volta's discovery of electricity, and of electricity's effects on the muscles of frogs, tended to further support the vision of the world as a space pervaded by fields of force.

The new discoveries were spread by means of the printing press. Reading, as McLuhan points out, necessarily induces people to rely more on their visual sense. A subtler effect of printed books is that the same book can be present in many places; this too enhanced the view of the world as space.

We find the mathematicians of this era concerning themselves with many kinds of space forms that earlier thinkers had deemed to be intractable. Instead of ignoring shapes that lacked simple names, the Renaissance mathematicians sought to understand them as actual forms.

3. Industrial Revolution. A complex machine is a perfect example of a logical system. Chains of cause and effect dart back and forth through the machine with wholly predictable outcomes. What could be more logical than a steam engine? This was the Age of Logic.

Newton's classic *Principia* is essentially a logical system. Instead of worrying about how objects could "act at a distance" and exert forces on each other, Newton chose to find universal laws to describe the action of the world's machinery. For Newton, the solar system is like a huge clock. The full flowering of the logic-based approach to physics came with Maxwell's equations for electricity and magnetism. Even though Maxwell stated his equations in terms of space-like force fields, he is known to have thought of them as descriptions of invisible machinery.

Economists of the Industrial Revolution liked to think of society as a big machine. The notion of an "economic man" who acts rationally to maximize his wealth is wholly logical and wholly mechanistic.

In mathematics, the Industrial Revolution saw algebraic calculation rise to new levels of expertise. George Boole had the notion of converting rational thought into a kind of algebra, and symbolic logic was born.

4. Modern Times. One thing that characterizes the modern era is the loss of certainty. Truth is no longer some numberlike rules, nor a serene space-like picture, nor a system of logic. Truth is, rather, infinitely complex. The crackling voice of the radio speaks as if from the endless ether beyond, as it should in the Age of Infinity.

Statistical mechanics explain such phenomena as heat and pressure in terms of large numbers of interacting individuals. The numbers of atoms involved are, relative to us, infinite. These figurative infinities become actual in quantum mechanics. The essence of modern quantum mechanics is that *no complete description of the world* can be contemplated; due to the uncertainty principle, quantum mechanics must treat the world as something that is essentially beyond our full comprehension. Mathematically, the theory is presented in terms of infinite-dimensional space. Relativity theory makes possible the science of cosmology, which attempts to study the universe as a whole.

In society, the modern era has seen the rise of mass movements and world war. This has led to widespread alienation, as people despair of having any effect on the essentially infinite complexities of modern life. Many individuals have turned aside from society to seek a direct mystical union with the Infinite itself.

In mathematics, the modern age brought Georg Cantor's set theory, an exact science of the infinite, and Godel's theorem set bounds to what logic can do without intuition.

5. The Postmodern Era. The computer is so suggestive a model for a human brain that it forces us to look at what it is we actually do. It is now quite normal to regard oneself as a finite system that processes information, and not as an immortal soul. The soaring cosmic uncertainties that inspired the Modern era can now be viewed as simple limitations on human information complexity.

Contemporary physicists are turning more and more to methods of computer simulation. An idea very much in the air these days is that physical systems can be thought of as information processors. A drop of ink is placed in a glass of water, and the water processes the drop into streamers. People are proving results about how rapidly a physical system can generate information and, working in a slightly different direction, some researchers are trying to show that certain simple physical systems behave like irreducibly complex "universal computers."

Socially, the rise of information is ending the alienation so characteristic of modern mass culture. Mass-produced products are being individualized — think of cable TV and of personalized automobile plates. Computerized polling procedures are giving the citizen new power over the government. The notion of information-theoretic complexity as an absolute good pervades postmodern ethics, leading to ever-increasing pluralism.

In mathematics, the study of information is leading to a number of fascinating new insights. *Mind Tools* is designed to give you a feeling for information, and to show you how information relates to number, space, logic, and infinity.

1

NUMBER

Zero and One

For one reason or another, we humans very commonly use 0 and 1 as an example of the most basic possible type of distinction. We think in terms of many other such distinctions — off and on, negative and positive, night and day, woman and man — but, at least to a mathematician, zero and one have a special appeal.

What is it about zero and one? Suppose I free-associate a little.

I think it is significant that the international hand gesture for coitus consists of a forefinger (1) bustling in a thumb and forefinger loop (0). This key archetype finds socially acceptable expression as the fat lady and the thin man. The symbol 0 seems egg-like, female, while 1 is spermlike and male. Can this really be an accident? A study of the history of mathematics shows that our present-day symbols were only adopted after centuries of trial and error. It seems likely that the symbols to survive are those that best "fit" our ingrained modes of thought.

An egg is round, and a sperm is skinny. There's only one egg, but there are lots and lots of sperm. The sperm are active, while the egg just waits. The lovely round woman is besieged by suitors seeking rest, seeking completion.

Formally speaking, both zero and one are undefinable. The primitive concepts "nothing" and "something" cannot be explained in terms of anything simpler. We understand these concepts only because they are built into the world we live in.

Before the beginning of time, the universe is like a 0, an egg. The egg hatches, and out comes 1, the chicken. Chickens here, chickens there, chickens, chickens everywhere.

Why, incidentally, does the egg hatch? Why is there something instead of nothing? Well, if there *weren't* any somethings, then we wouldn't be here asking about it. . . . Is this enough of an answer?

Occasionally, in moments of deep absorption, all distinctions may seem to fall away, and you do have a kind of 0 experience. As long as you are, in some sense, in touch with the void, you aren't going to be asking, "Why is there something instead of nothing?" But the moment passes, and you're a regular person again. Cluck, cluck.

One and zero are like Punch and Judy, endlessly acting their play. What is perhaps astonishing is that mankind has managed to build up a science based on this play: the science of number.

Numbers and Logs

The simplest possible numeration system is the "tally" system. Here a number N is represented by a list of N ones. Ordinary finger-counting uses this system (Fig. 20).

The tally system is awkward for expressing big numbers. The Roman numeration system is considerably more efficient. Recall that the Romans used I, V, X, L, C, D, and M to stand for, respectively, 1, 5, 10, 50, 100, 500, and 1000. Writing X instead of 1111111111 is a big savings. One flaw with the Roman system is that you can't always read a Roman number one letter at a time — to know that IV is 4 and LX is sixty, you have to see two letters together. This makes Roman numbers hard to read. How many of us have seen the Roman number MDCCLXXVI on the back of a dollar bill without realizing that it's supposed to be 1776? Another problem with the Roman system is that it peters out at M, so that if I want to express eleven thousand, I have to write MMMMMMMMMMM. Out past M, something like the tally system sets in again.

These days, every advanced culture on Earth uses the familiar base-ten system of numeration. Quantities are expressed as sums of units, groups of ten, groups of ten tens, groups of ten groups of ten tens, and so on:

39_{TEN} = 3 groups of ten and 9 units
322_{TEN} = 3 groups of ten tens, 2 groups of ten, and 2 units
1987_{TEN} = 1 group of ten groups of ten tens, 9 groups of ten tens, 8 groups of ten, and 7 units.

The subscript "TEN" is a reminder that our number system is based on groupings of ten. It is a little hard to grasp that there is nothing sacred about the base-ten numeration system. Simple-minded as it may seem, the only reason we use ten is that we have ten fingers. A numeration system could equally well be based on, say, three — with groups of three, groups of three threes, groups of three groups of

BASE TEN	HAND	TALLY
1		I
2		I I
3		I I I
4		I I I I
5		I I I I I

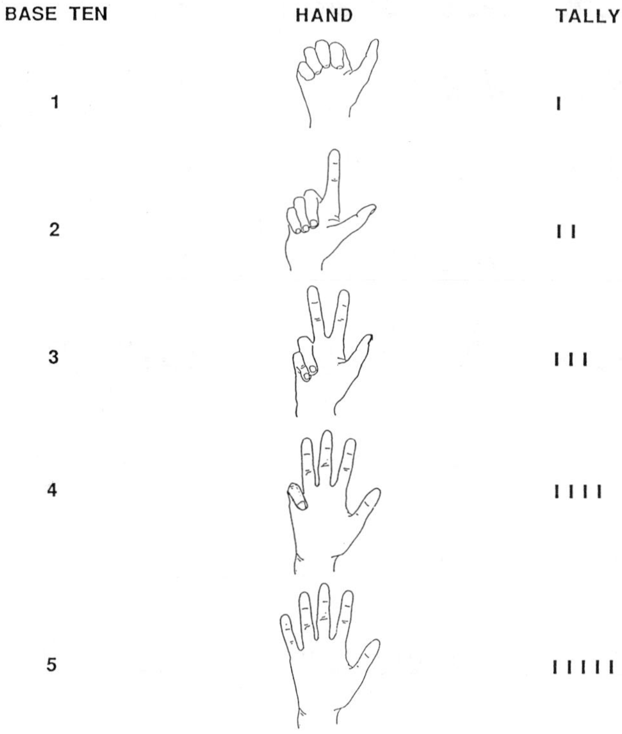

Fig. 20 Counting to five.

three threes, and so on. To make this clear, take a look at the different ways in which 54 units are broken up and described in numeration systems based on two through ten.

Digital computers use the "binary" or base-two number system. One reason for this is that computer memories are based on internal switches which can be set two ways: *on* or *off*. By letting on be 1 and off be 0, computers can represent numbers by setting their memory switches in various arrangements.

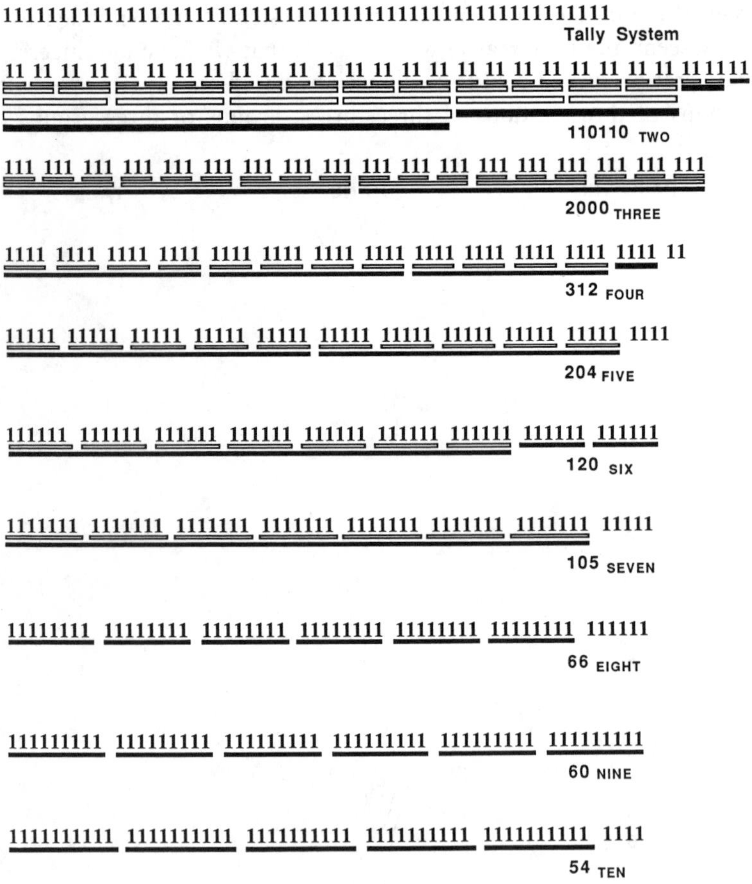

111111111111111111111111111111111111111111111111111111
Tally System

11 11 11 11 11 11 11 11 11 11 11 11 11 11 11 11 11 11 11 11 11 11 11 11 111111
110110 TWO

111 111 111 111 111 111 111 111 111 111 111 111 111 111 111 111 111 111
2000 THREE

1111 1111 1111 1111 1111 1111 1111 1111 1111 1111 1111 1111 1111 11
312 FOUR

11111 11111 11111 11111 11111 11111 11111 11111 11111 11111 1111
204 FIVE

111111 111111 111111 111111 111111 111111 111111 111111 111111
120 SIX

1111111 1111111 1111111 1111111 1111111 1111111 1111111 11111
105 SEVEN

11111111 11111111 11111111 11111111 11111111 11111111 111111
66 EIGHT

111111111 111111111 111111111 111111111 111111111 111111111
60 NINE

1111111111 1111111111 1111111111 1111111111 1111111111 1111
54 TEN

Fig. 21 54 in different number bases.

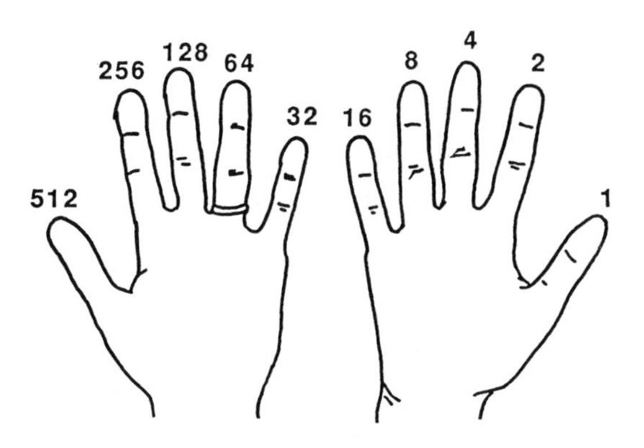

Fig. 22 Fingers as base-two flip-flops.

By applying a binary system to the positions of your ten fingers, you can actually count up to 1,023. This is done as follows. Lay your two hands down side by side, palms up as in Fig. 22. Running from right to left, associate each digit with a successive number from the "doubling sequence": 1, 2, 4, 8, 16, 32, 64, 128, 256, and 512. These are, of course, powers of two. The "grouping by twos" process suggested in Fig. 21 indicates that each whole number can be represented in a unique way as a sum of some numbers from the doubling sequence.

If you stick up the appropriate finger for each doubling-sequence number used in a given number's breakdown, you get a finger pattern specific to that number, as illustrated in Fig. 23. To the right of the finger patterns, I have written the corresponding base-two notation for each pattern.

Using only one hand, you can represent any number between 00000_{TWO} and 11111_{TWO}, which is any number between 0 and 31_{TEN}; 32_{TEN} possibilities in all. If you use your toes as well as your hands (though raising and lowering toes is not easy) you could reach $11111111111111111111_{\text{TWO}}$, which is $1{,}048{,}576_{\text{TEN}}$, or a bit more than a million.

If we try to base a numeration system on one, we end up with the tally system, because all powers of 1 are equal to 1. Two then has the virtue of being the smallest number that can be used as the base of a numeration system. For purposes of analyzing how much information a decision among *N* possibilities involves, it is good to think in terms of base two.

To choose among thirty-two possibilities is to make five 0/1 choices in writing out a five-digit binary number. Another way of putting this is to say that a choice among thirty-two possibilities requires five bits

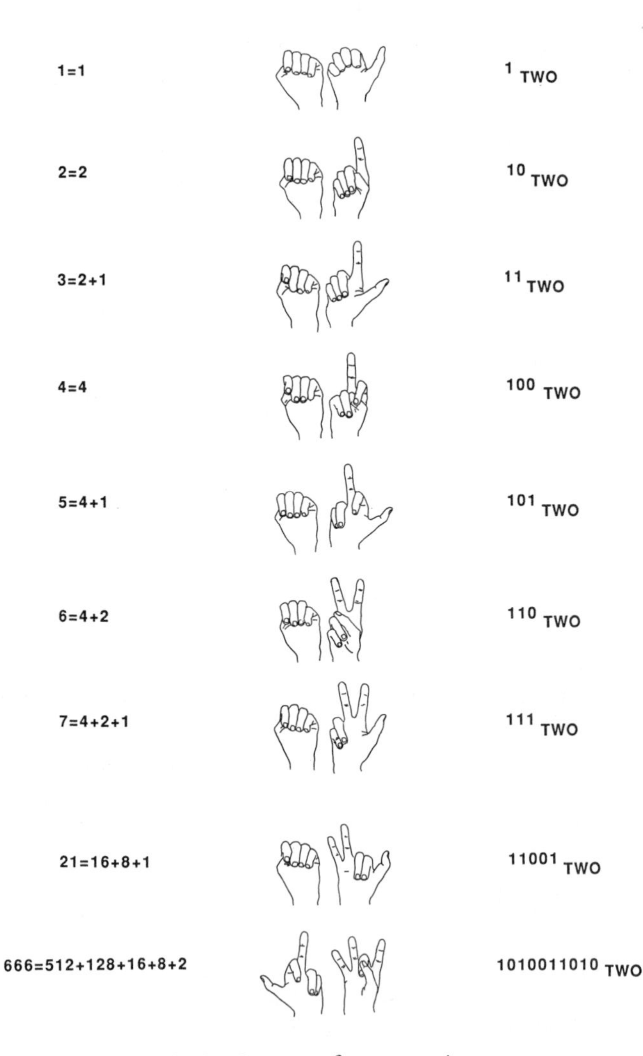

Fig. 23 Base-two finger counting.

of information, where a bit of information is a single choice between 0 and 1. It is significant that thirty-two is two to the *fifth* power, and a choice among thirty-two possibilities takes *five* bits of information. In general, choosing among two to the N possibilities requires N bits of information. Making the choice is like selecting a path along a path that has N binary forks in it.

To express this insight more clearly, we need to talk about logarithms. Information theory is usually formulated in terms of logarithms. In the past, this always put me off; like anyone else, I hated logarithms, but then I had a key insight: *the logarithm of a number is approximately equal to the number of digits it takes to write the number out.* In base ten, log 10 is 1, log 100 is 2, log 1987 is about 3, log 12345 is about 5, and the log of one billion is 10.

Formally, we define the logarithm operation as the inverse of exponentiation. In base ten, the logarithm of a number N is precisely the power to which ten would have to be raised to equal N.

Logarithm	is to	exponent as
division	is to	multiplication, as
subtraction	is to	addition.

$y = \log_{10} x$	means that	$10^y = x$.
$y = x/10$	means that	$10y = x$.
$y = x - 10$	means that	$10 + y = x$.

We don't have to base our logarithms on ten. When we are doing information theory, it is much better to base our logarithms on two. In base two, the logarithm of a number N is precisely the power to which two would have to be raised to equal N:

$$y = \log_2 x \text{ means that } 2^y = x.$$

No matter what base we use, logarithms have a number of very beautiful properties. Three properties that make logarithms useful for calculation are

$$\log x^N = N \log x;$$
$$\log xy = \log x + \log y;$$
$$\log x/y = \log x - \log y.$$

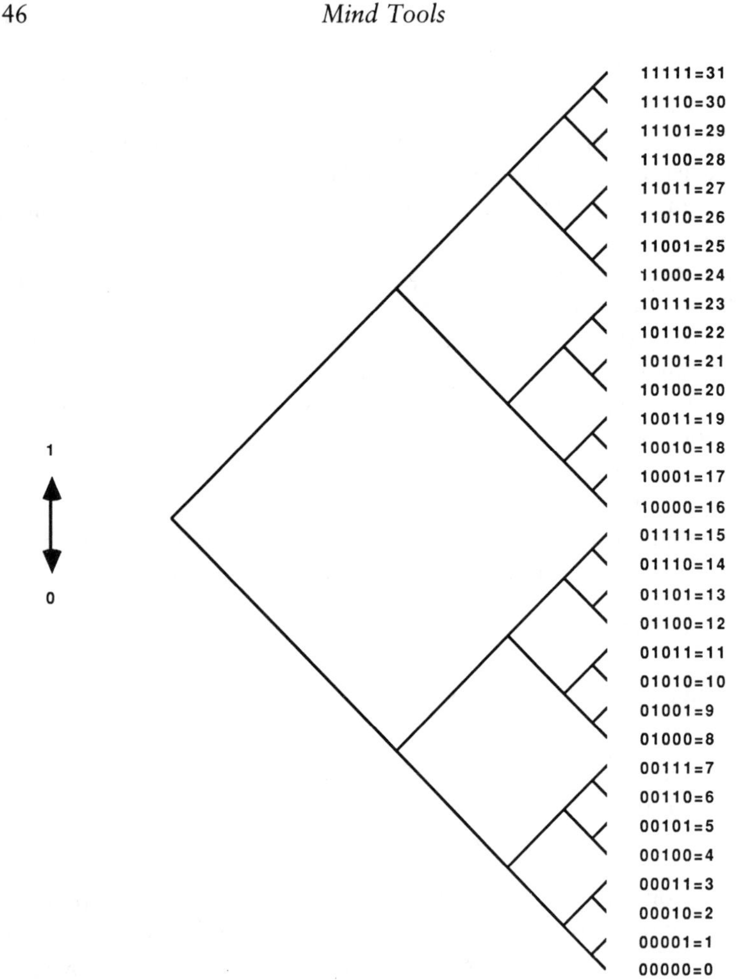

Fig. 24 Five 0/1 choices give thirty-two outcomes.

In the old days, people used to compute a large product like 1987 × 7891 by looking up log 1987 and log 7891, adding the logs, and looking up the number whose log is equal to the sum. The computation was done either by means of a log-antilog table, or by means of a slide rule. A slide rule basically consists of two identical sticks printed with number names in such a way that a number's name appears at a position whose actual distance from the stick's end is the logarithm of the number. Products are computed by the analog process of measuring off the corresponding stick lengths. Electronic calculators actually do their work by using logarithms, though one doesn't notice this except when the hidden log-antilog steps lead to round-off errors such as saying "1.9999999" when "2" is meant.

It is often useful to draw graphs in which one or both axes are scaled logarithmically. Looking at the graphs on page 48 makes it clear that, working in any given base, the number of digits it takes to write a number N lies between $\log N$ and $(\log N) + 1$.

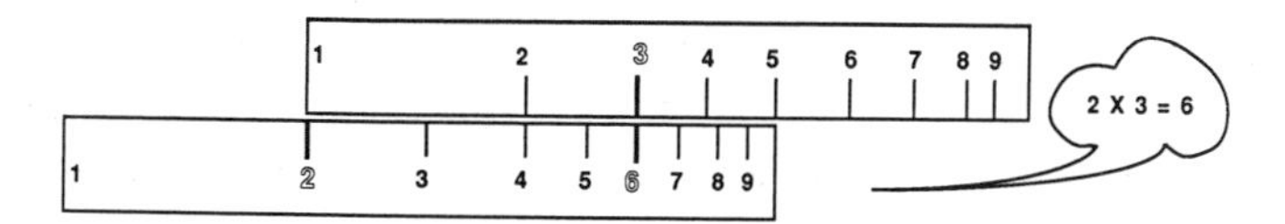

Fig. 25 A slide rule in action.

The reason information theory makes such frequent use of logarithms base two is that the most fundamental unit of information is a bit, and a bit of information represents a choice between two possibilities. The number of digital bits it takes to write out a number N in binary notation is approximately equal to the base-two logarithm of N.

Shannon thought of information in terms of *reduction of uncertainty*. If someone is about to send you a message, you are uncertain about what the message will say, and the greater the number of possible messages, the greater your uncertainty. The arrival of the message reduces your uncertainty and gives you information. Shannon defines the information in the message as equal to the base-two logarithm of the total number of messages that might possibly have been sent. If we write I for his measure of information and M for the number of possible messages, then we can write Shannon's definition as an equation:

$$I = \log_2 M.$$

This means that if there are only two possible messages, say "Yes" and "No," then getting the message gives you one bit of information.

Four possible messages means two bits of information, eight possible messages means three bits of information, and so on. One million is approximately two to the twentieth power, so if there are one million possible messages, each message carries some twenty bits of infor-

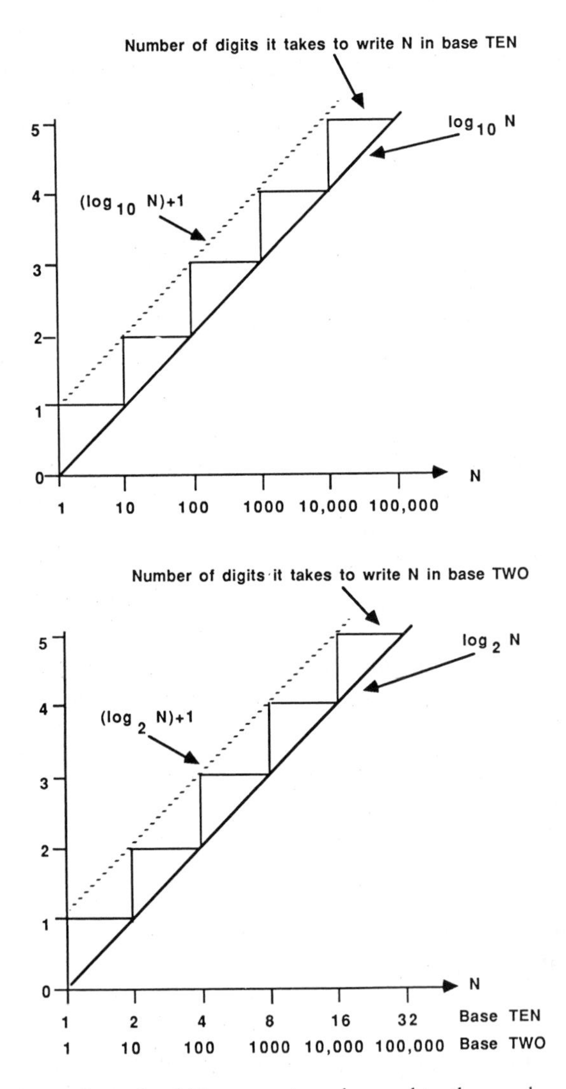

Fig. 26 In any base, log *N* is approximately equal to the number of digits it takes to write *N*.

mation. Put differently, one can choose among one million possibilities by asking twenty yes-or-no questions.

In general, writing a number in binary notation takes about three times as many digits as it does to write the number in decimal notation. Nevertheless, we do not think of decimal notation as a more efficient code for a number, because the information cost of remembering a decimal digit is about three times the cost of remembering a binary digit. This is made quite precise by an equation that relates different-based logarithms:

$$\log_2 M = (\log_2 10)\,(\log_{10} M).$$

Think of $\log_2 M$ as the number of binary symbols it takes to express M; think of $\log_{10} M$ as the number of decimal symbols it takes to express M; and think of $\log_2 10$ as a conversion factor giving the number of binary bits it takes to choose one of the ten decimal symbols: 0, 1, 2, 3, 4, 5, 6, 7, 8, 9. Put differently, $\log_2 10$ is the information content per decimal digit. The information in M is the product of the information per digit and the number of digits it takes to write M. The value of $\log_2 10$ is, by the way, 3.32. That means a decimal digit is worth 3.32 bits of information. $\text{Log}_2\ 26$ is 4.70, which means that a letter of the alphabet is worth 4.7 bits of information. (See Fig. 27.)

Number Patterns

Whole numbers have an obstinate solidity to them. No matter how much you try, you cannot break 5 into two equal whole numbers — 5 breaks into a 1 and a 4, or into a 2 and a 3, and that's all there is to it. Someone might say, "What about breaking 5 into 2½ and 2½?" but for now we're not interested in that kind of talk. Right now we're thinking of numbers as made up of unbreakable units.

A number is an eternal, universal pattern. The Greeks may have identified certain of their gods with numbers, but now those gods are forgotten, and the numbers live on. It is perhaps conceivable that some extraterrestrial races do not think in terms of number at all. Sentient clouds of gas, for instance, might never encounter any sensations discrete enough to suggest the idea of counting. But if any intelligent race does form the idea of discrete objects, we may be certain that they will arrive at just the same numbers that we use. Number is an objectively existing feature of the world.

Different numbers of units make up various kinds of patterns. In the Introduction we talked at some length about how certain archetypes are associated with some of the smaller numbers:

1 monad — unity
2 dyad — opposition
3 triad — thesis–antithesis–synthesis
4 tetrad — balance of quaternity
5 quintad — a step further out

Now let's go on and look at some other notions that various familiar numbers suggest. In doing this, I will often refer to the Pythagoreans. Pythagoras himself lived in Greece and Italy during the sixth century B.C. He was not only a philosopher, but something of a wizard. His followers banded together, to form a kind of religious community on the island of Samos. The most influential doctrine of these "Pythagoreans" was the notion that, at the deepest level, the universe is made of numbers. In his *Metaphysics,* Aristotle describes the Pythagorean teaching:

> They thought they found in numbers, more than in fire, earth, or water, many resemblances to things which are and become; thus such and such an attribute of numbers is justice, another is soul and mind, another is opportunity, and so on; and again they saw in numbers the attributes and ratios of the musical scales. Since, then, all other things seemed in their whole nature to be assimilated to numbers, while numbers seemed to be the first things in the whole of nature, they supposed the elements of numbers to be the elements of all things, and the whole heaven to be a musical scale and a number.

The Pythagoreans thought of numbers as embodying various basic concepts. In general, the even numbers (2, 4, 6, 8, . . .) were thought of as female, and the odd numbers (3, 5, 7, 9, . . .) were thought of as male. Two specifically stood for woman, and 3 for man. One reason for this may lie in the appearance of the male and female genitalia.

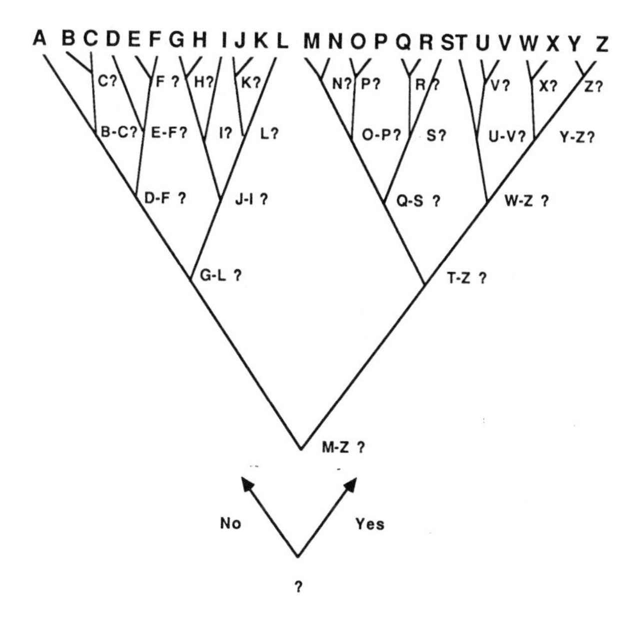

Fig. 27 A letter is worth a little less than five bits.

Given that 5 is the sum of 2 and 3, 5 was thought of as standing for marriage.

Another Pythagorean train of thought held 1 to stand for God, or the unified Divinity underlying the world. Two, as the first moving away from unity, stood for strife, or opinion. This interpretation of 2 is present in the notion of the dyad, which pairs opposing concepts.

Three is usually thought of as a lucky number; the Christian religion is based on the notion of a triune God. Another important feature of 3 is that three dots describe the simplest possible two-dimensional shape: the triangle. Churches in Austria and southern Germany usually have over their altars pictures of God as an eye inside a triangle. For some reason, this image has found its way onto the American dollar bill.

In the rest of this section I am going to be thinking of numbers primarily as patterns in space. That is, I will think of numbers as collections of dots, and will look at some of the patterns into which various-sized collections can be formed. The Pythagoreans often used space patterns to shed light on number properties. In doing this, one tries to keep the patterns as simple and as "digital" as possible; the primary topic here is, after all, number rather than space.

One of the earliest examples of people using dot patterns to represent numbers appears in the Chinese image known as the *lo-shu*. Here there are patterns representing the numbers 1 through 9, and these patterns are arranged into a "magic square." Supposedly, Emporer Yu saw the *lo-shu* pattern on the back of a tortoise on the banks of the Yellow River in 2200 B.C. The pattern is called a magic square because the sum of the numbers along any horizontal, vertical, or diagonal is always the same: 15.

My very long Fig. 29 includes simple pictures of as many of the numbers 1 through 100 as I could come up with. Some of the pictures are traditional, some are my inventions. Readers may be able to discover some simple patterns that I have overlooked entirely. As you read through the rest of this section's text, keep referring to Fig. 29.

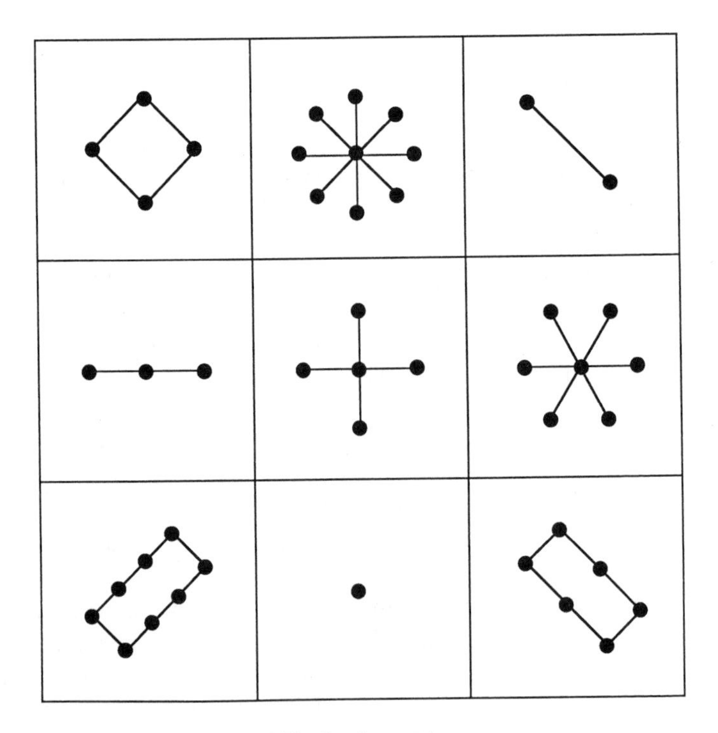

Fig. 28 The *lo-shu* magic square.

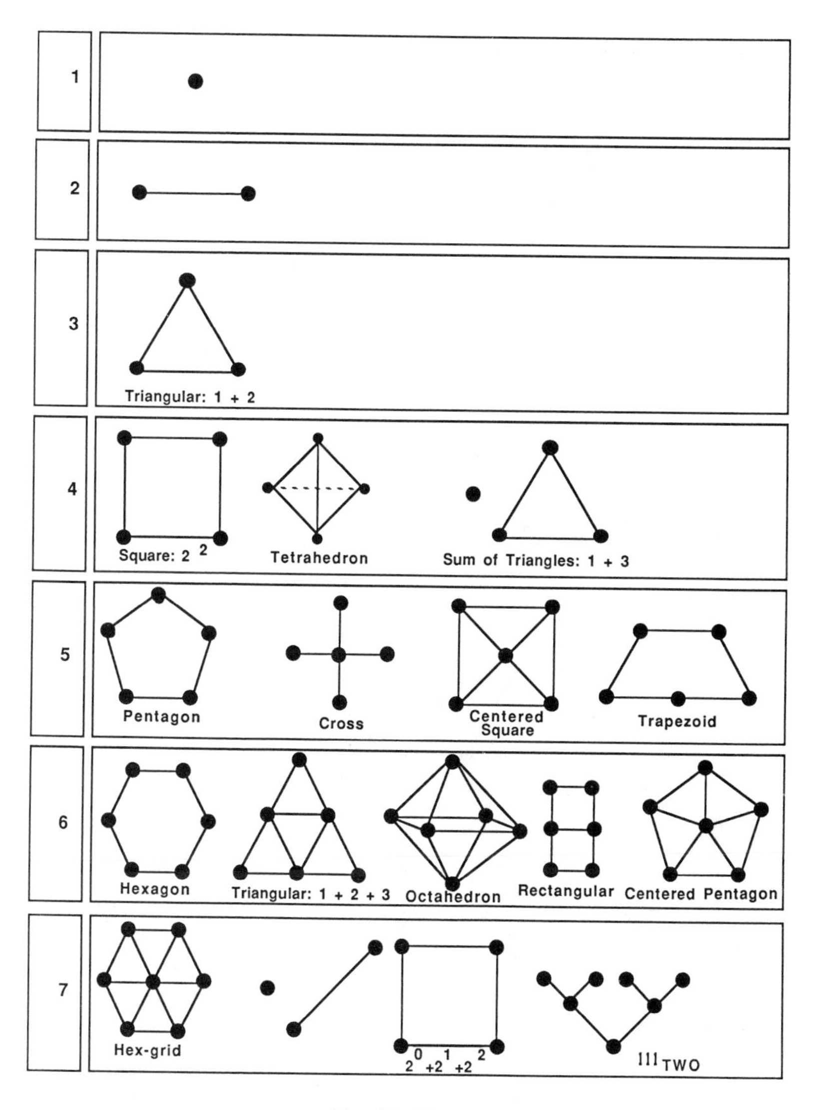

Fig. 29, No. 1.

Since 4 breaks so readily into the equal halves 2 and 2, the number 4 was thought to stand for justice. Some residue of the notion survives in the expression "a square deal." The vertical sides of the square make square (or rectangular) shapes ideal for building things. Perhaps it is this stability that has made "square" come to mean a solid (and

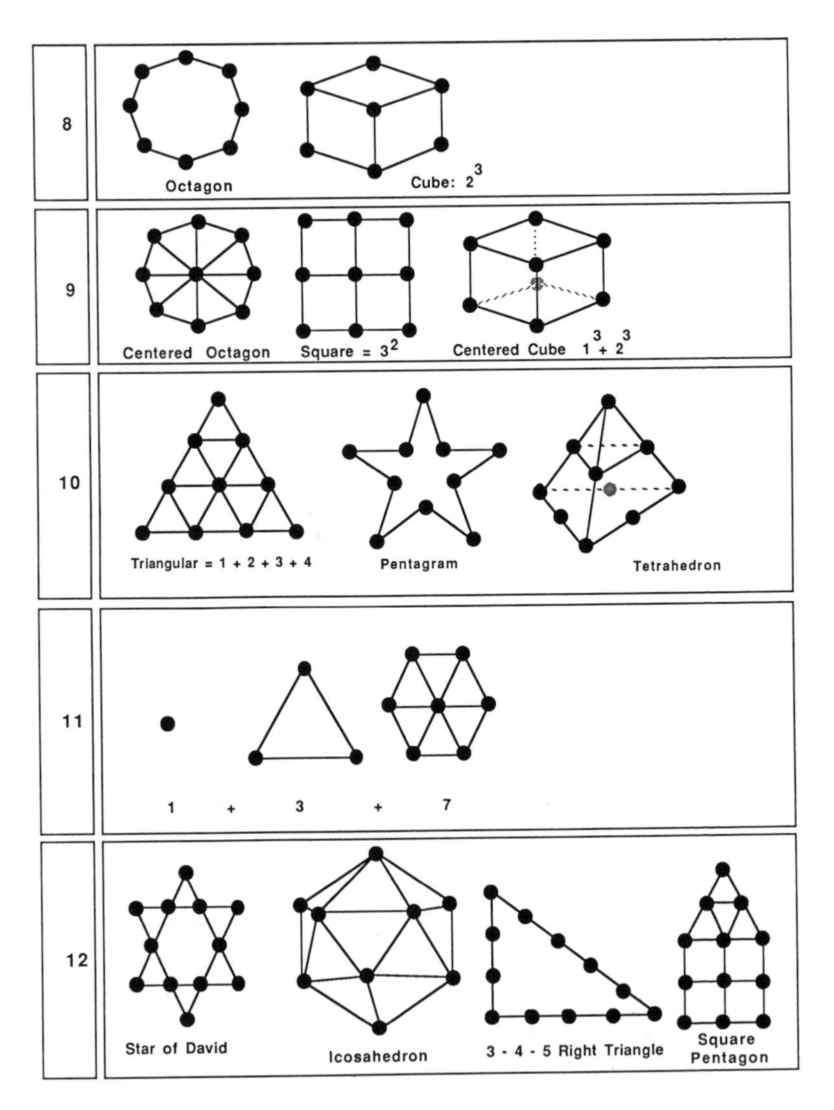

Fig. 29, No. 2.

somewhat boring) citizen. A unique feature of 4 is that not only is it 2 plus 2, it is also 2 times 2 and 2 to the 2nd power. At this low number level the distinctions between addition, multiplication, and exponentiation are not yet fully developed.

The number 5 is significant for many reasons. A human body has five big things sticking out of it (head, two arms, two legs), and the

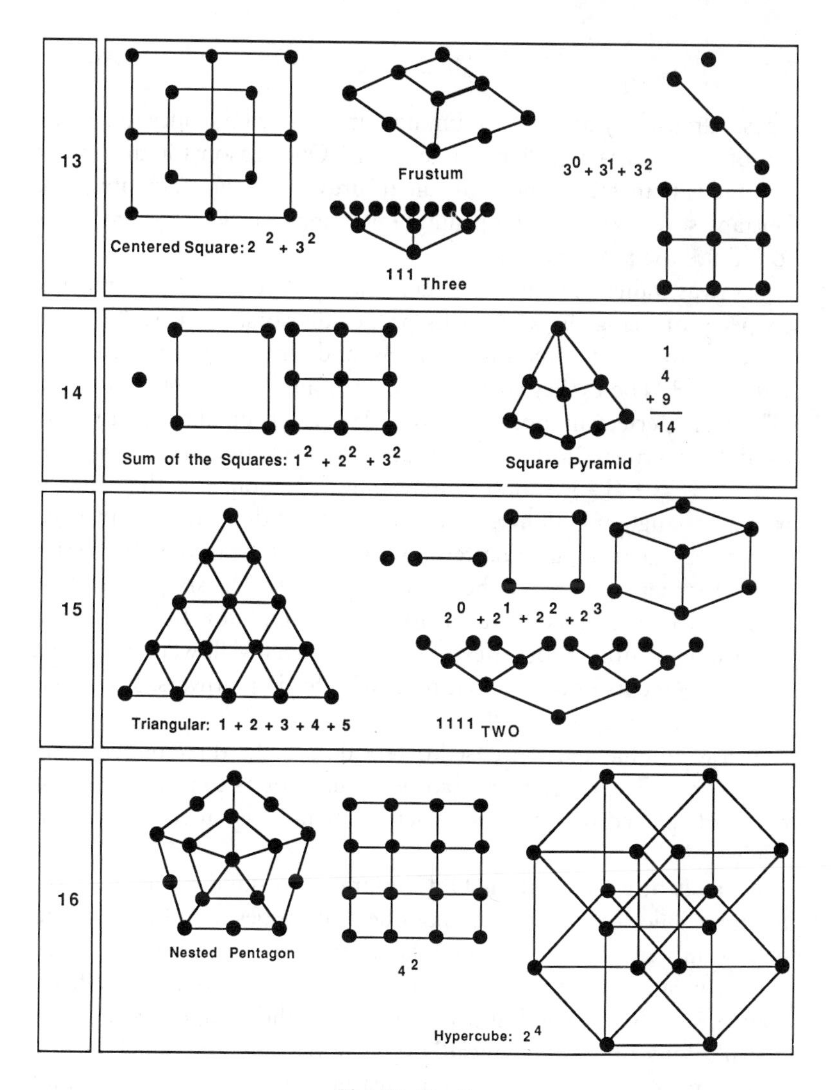

Fig. 29, No. 3.

Pythagoreans sometimes identified 5 with the body or with health. Another important human aspect of 5 is that an undamaged hand has five fingers. Followers of Pythagoras sometimes wore lucky five-pointed stars . . . just like sheriffs in a Western! Children usually draw houses as irregular pentagons.

Somehow 5 also acquired a negative significance during the Middle

Ages. Almost any attempt to summon up the devil includes the drawing of a pentagram (a five-pointed star). One reason for this might be that it is in fact rather difficult to draw an accurate pentagram. Perhaps sorcerers used the pentagram to make it hard for amateurs to imitate their ceremonies.

Six was valued by the Pythagoreans because it has the unusual property of being the sum of its proper divisors. That is, 6 can be broken into six 1s, three 2s, or two 3s, and remarkably enough, $6 = 1 + 2 + 3$. The Pythagoreans called such numbers *perfect* numbers. (The next perfect number after 6 is 28, which breaks into twenty-eight 1s, fourteen 2s, seven 4s, four 7s, or two 14s, and $28 = 1 + 2 + 4 + 7 + 14$.) Another interesting fact about 6 is that, like 3, it is a "triangular number," meaning that six dots can be arranged into a regular triangular pattern. Six is important in the Old Testament, for there God made the world in six days. The six-pointed Star of David came, in later years, to be a symbol for Judaism. Two interesting natural properties of 6 are that many flowers are based on arrangements of 6, and that bees, who feed on flowers, store their honey in six-sided wax cells.

A less-familiar pattern based on six dots is the three-dimensional *octahedron*. The octahedron can be thought of as two square-based pyramids placed base to base. Each of its faces is an equilateral triangle.

Seven is often used in the Old Testament to stand for a very large number. One reason for this might be that 7 does not allow itself to be put into any simple, symmetrical arrangement. Seven is said to be *prime,* meaning that it has no divisors other than 1 and itself. Of course 5 is also prime, but 5 has so many other properties that one doesn't think of it as a "typical" prime. Seven could be called the first typical prime. Another refractory feature of 7 is that, if we limit ourselves to the simple tools of ruler and compass, it is impossible to construct a regular seven-sided figure (heptagon). All the numbers less than 7 can be put into familiar patterns, but 7 is most naturally thought of as a straight line of dots, which could be why 7 suggests the idea of largeness, or of continuing on toward infinity. The uniqueness of 7 has led people to think of it as a lucky number; this association could also have something to do with the dice game of craps, where an initial roll of 7 wins automatically. Seven has the attractive property $7 = 111_{\text{TWO}} = 1 + 2 + 4$.

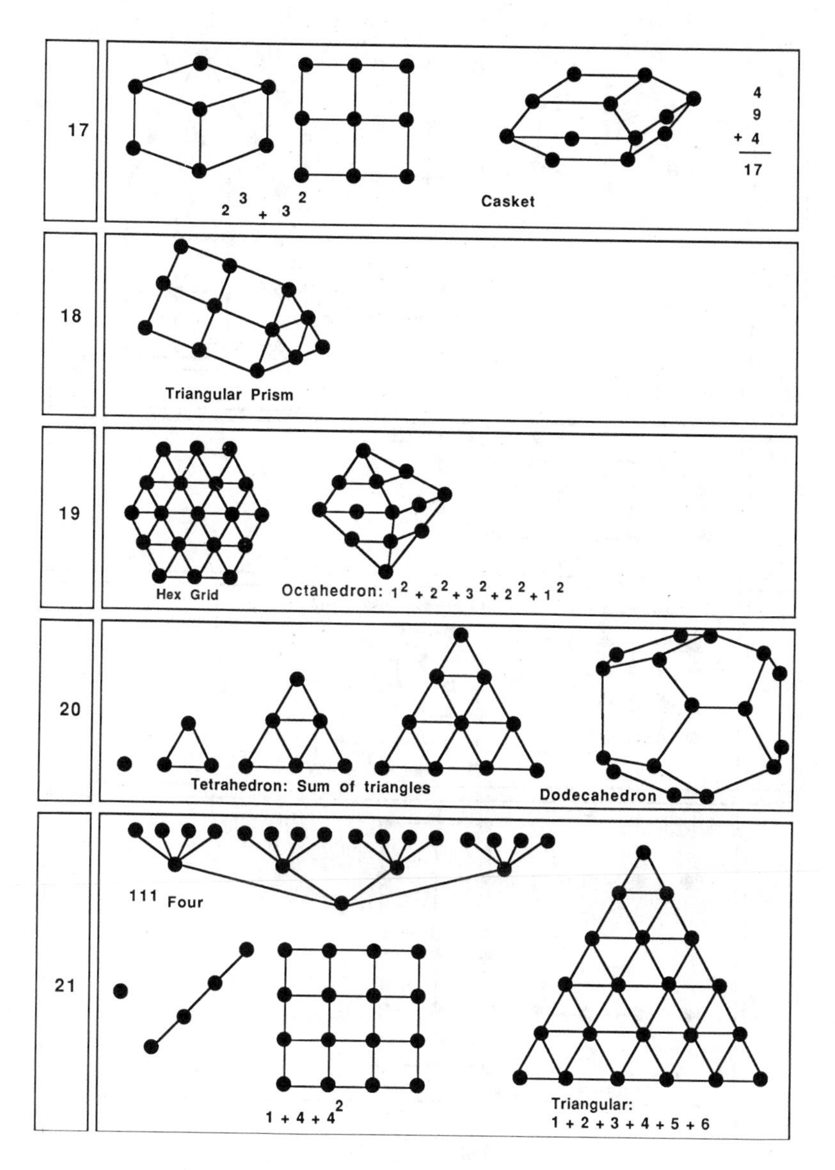

Fig. 29, No. 4.

Numbers that can be written as a series of 1s in some numeration system are called "rep-numbers." The nice thing about rep-numbers is that they can be drawn as tree patterns where each fork has the same number of branches. One other pattern that 7 embodies is that we can arrange seven dots into a hexagonal grid.

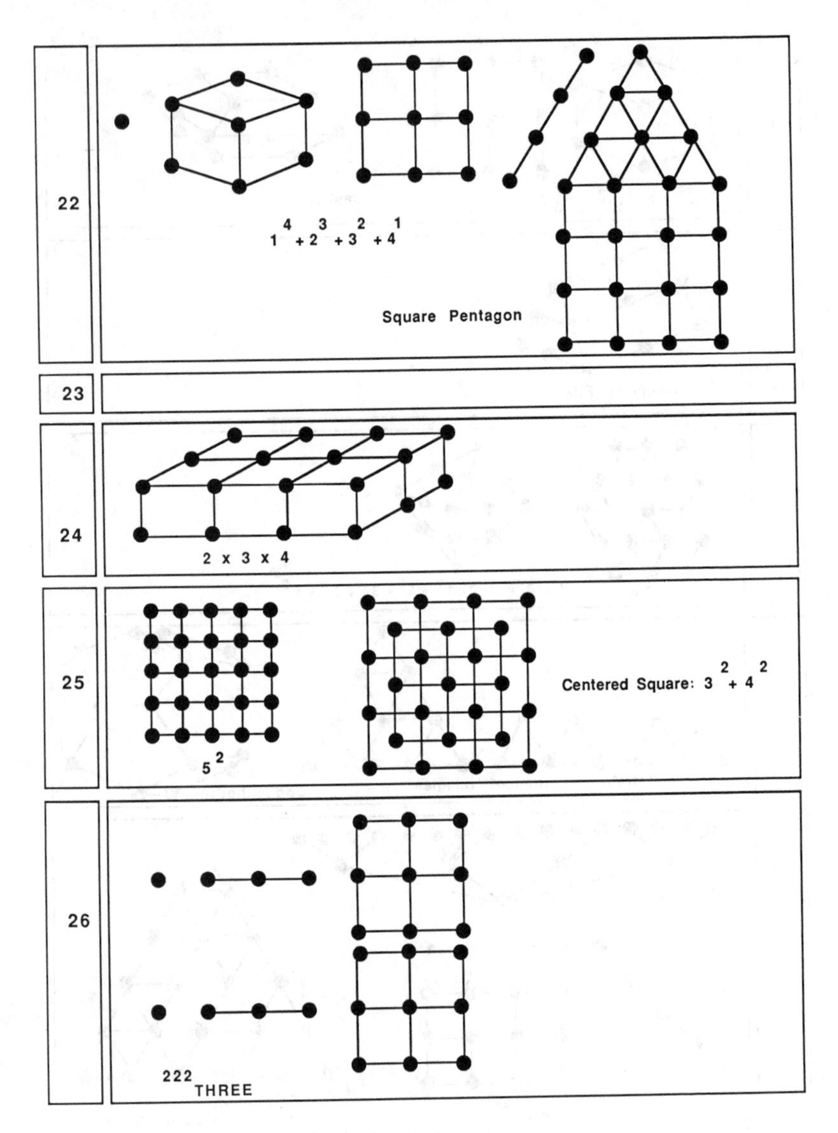

Fig. 29, No. 5.

Eight is an interesting number, since it is the first number that is a perfect cube — that is, $8 = 2 \times 2 \times 2$. Eight is, so to say, even with a *vengeance,* and its smooth curves make it really an even more feminine number than 2. A curious thing about our symbol for 8 is that, if we push this symbol onto its side, we get the common symbol for infinity. Perhaps it is 8's ability to be halved, rehalved, and halved again that suggests the notion of endlessness.

Nine is the first really typical square number. If we think of it as "thrice three," it sounds quite grand. An interesting thing about the transition from 8 to 9 is that one is going from two cubed to three squared! It is quite hard to think of nine dots without arranging them into a 3-by-3 grid; at 9 the human imagination is worn out, and we stop inventing new number symbols. One other way of thinking about

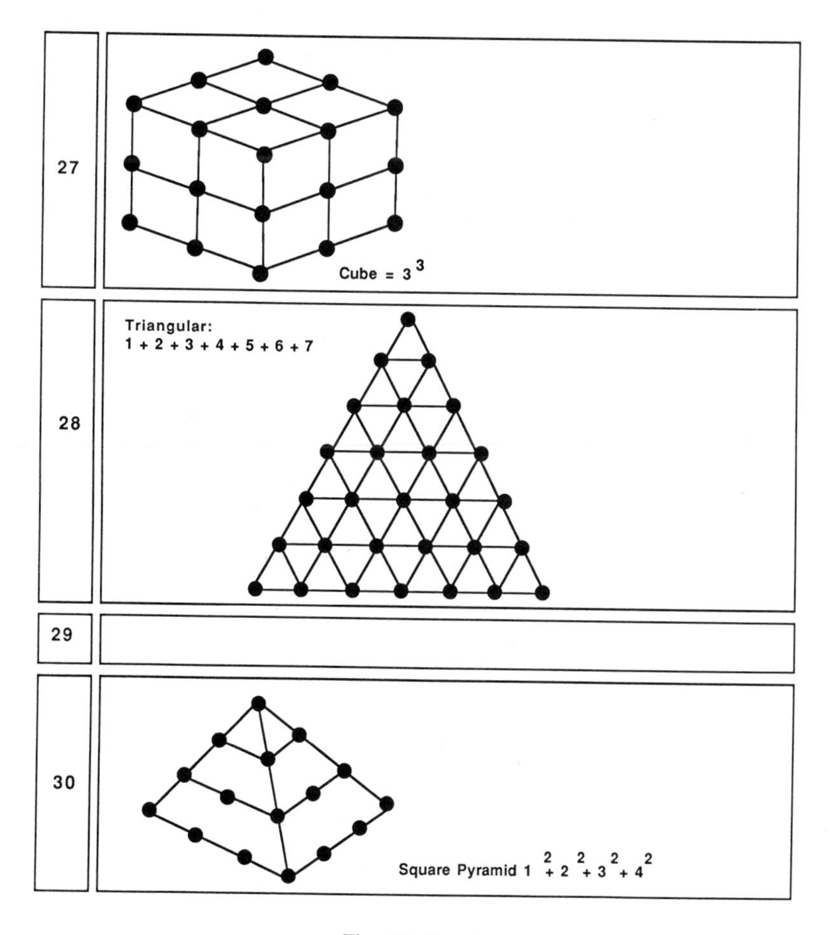

Fig. 29, No. 6.

nine is to imagine placing an extra dot in the center of a 2 × 2 × 2 cube to get a "centered cube."

Ten is where we begin using our positional decimal notation. Like 3 and 6, 10 is a triangular number: 10 = 1 + 2 + 3 + 4. (In general, any number is called triangular if it is the sum of a consecutive number sequence starting with 1.) This fact is, of course, the basis for the game of tenpins, or bowling. The Pythagoreans valued 10 highly, as 1, 2, 3, and 4 seemed so fundamental to them.

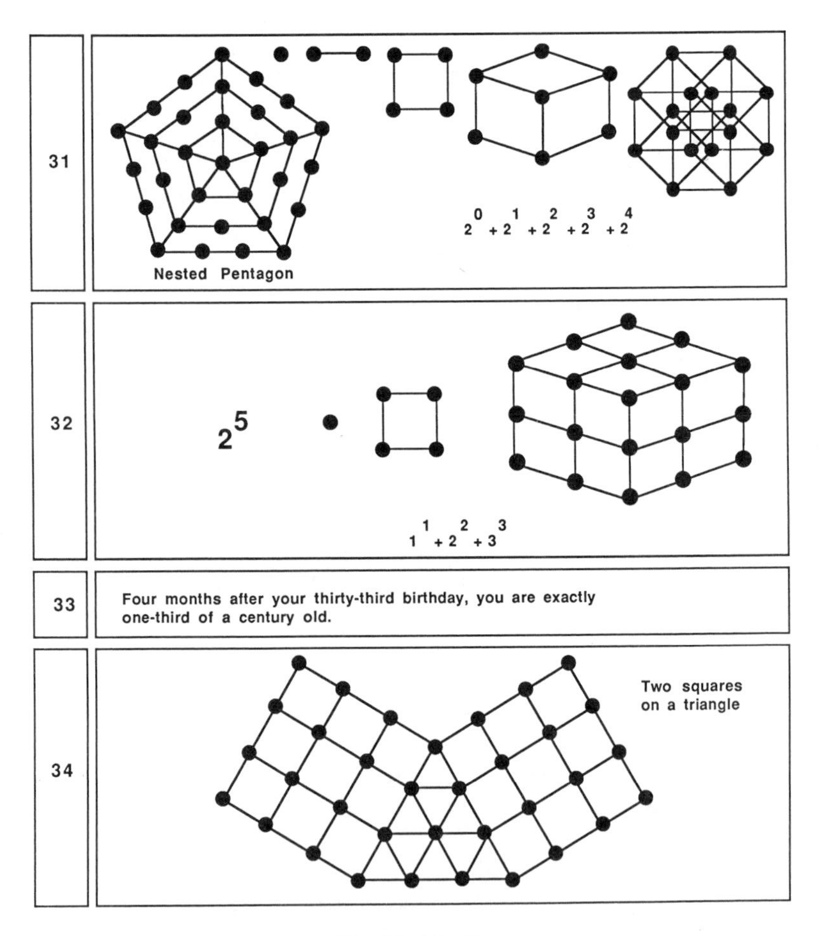

Fig. 29, No. 7.

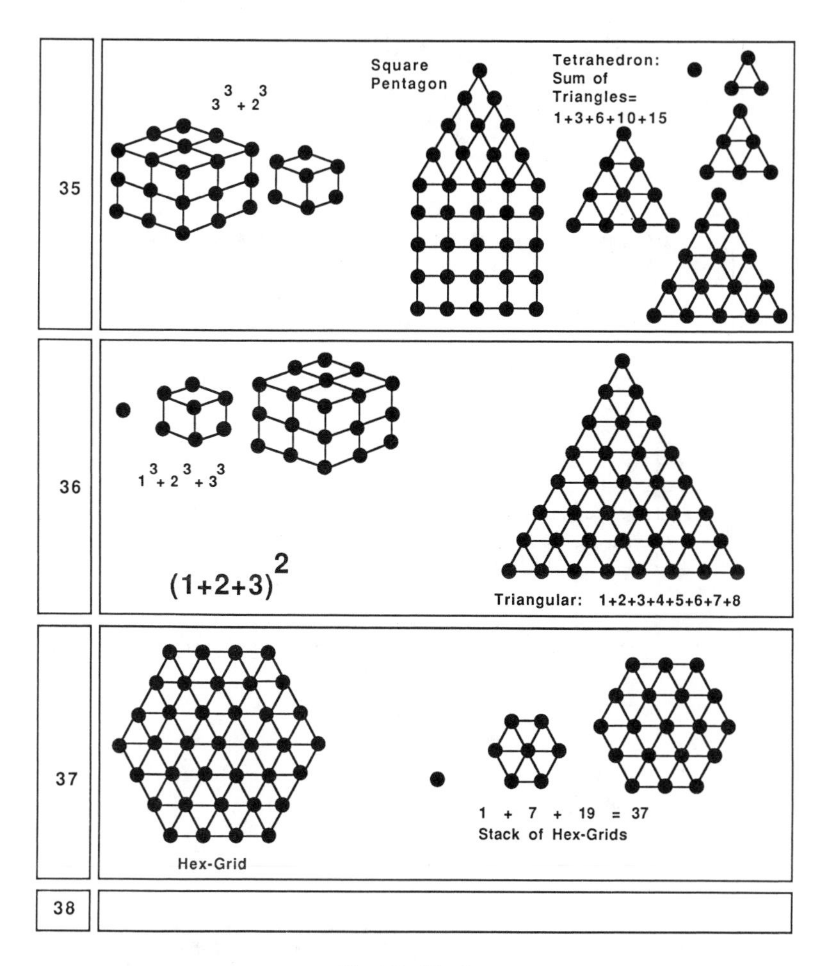

Fig. 29, No. 8.

Eleven is often thought of as a lucky number. This could be because, in our number system, 11 represents a new beginning, a fresh run through the basic sequence 1 to 9. As such, 11 makes one think of rebirth and increase. Another pleasant feature of 11 is the fact that its two digits are the same. This suggests notions of duplication and plenitude. It is also interesting that 11 is the sum of the three most magical earlier numbers: 1, 3 and 7.

Twelve is a number with very many associations. There were twelve apostles; eggs are sold in dozens; a clock shows twelve hours; a year

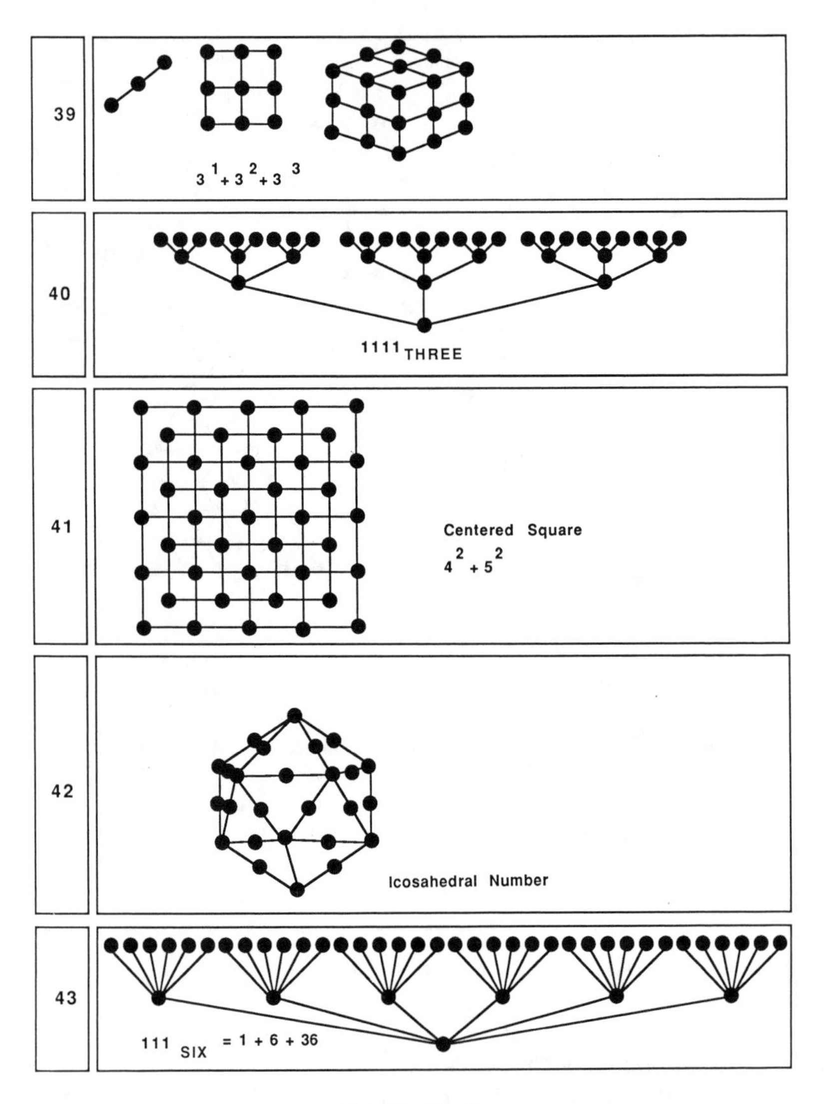

Fig. 29, No. 9.

has twelve months; and the zodiac has twelve signs. Twelve is "more divisible" than any larger number — that is, no number after 12 is divisible by so great a proportion of the numbers less than itself.

Twelve also has the property that, if we set a triangular pattern of six dots on top of a square pattern of nine dots (with one edge overlapping), we get a kind of "square pentagon" of twelve dots.

It turns out that there are exactly five "regular polyhedra" in three-dimensional space. A regular polyhedron is a solid that has two properties: (1) each corner looks the same as all the others, (2) each face is a regular polygon congruent to all the other faces. The five regular polyhedra, also known as the Platonic solids, are as follows:

tetrahedron	a four-cornered arrangement of four triangles
cube	an eight-cornered arrangement of six squares
octahedron	a six-cornered arrangement of eight triangles
icosahedron	a twelve-cornered arrangement of twenty triangles
dodecahedron	a twenty-cornered arrangement of twelve pentagons

Thirteen is thought of as an unlucky number. This could be because, in the New Testament, Judas is the thirteenth apostle. The notion of Friday the thirteenth being unlucky is probably related to Judas and to the Good Friday crucifixion. Another unsettling thing about 13 is that it is prime, and even more so than 7, lacking in any interesting structural properties. Another reason for 13's unpopularity is that it contains the second, and thus perhaps *false,* appearance of the divine number 3. Still, if we scratch beneath the surface, we find that 13 has one very interesting feature: $13 = 9 + 3 + 1 = 111_{\text{THREE}}$.

Another pattern which can be formed with 13 dots is a "centered square." The centered squares can be thought of as patterns of nested squares. Five, although we did not mention this before, is also a centered-square number.

Fourteen, being twice 7, is a fairly lucky-seeming number. A nice thing about 14 is that it is the sum of three perfect squares: $14 = 1^2 + 2^2 + 3^2$. This enables us to think of 14 dots as arranged into a kind of pyramid pattern.

Fifteen is another triangular number: $15 = 1 + 2 + 3 + 4 + 5$. Fifteen is also a base-two rep-number and can therefore be drawn as a tree pattern.

Sixteen is the first number that is the fourth power of a number; that is, 16 is 2 raised to the 4th power. Thus 16 suggests the mysterious notion of four-dimensional space. It is possible to draw a sixteen-

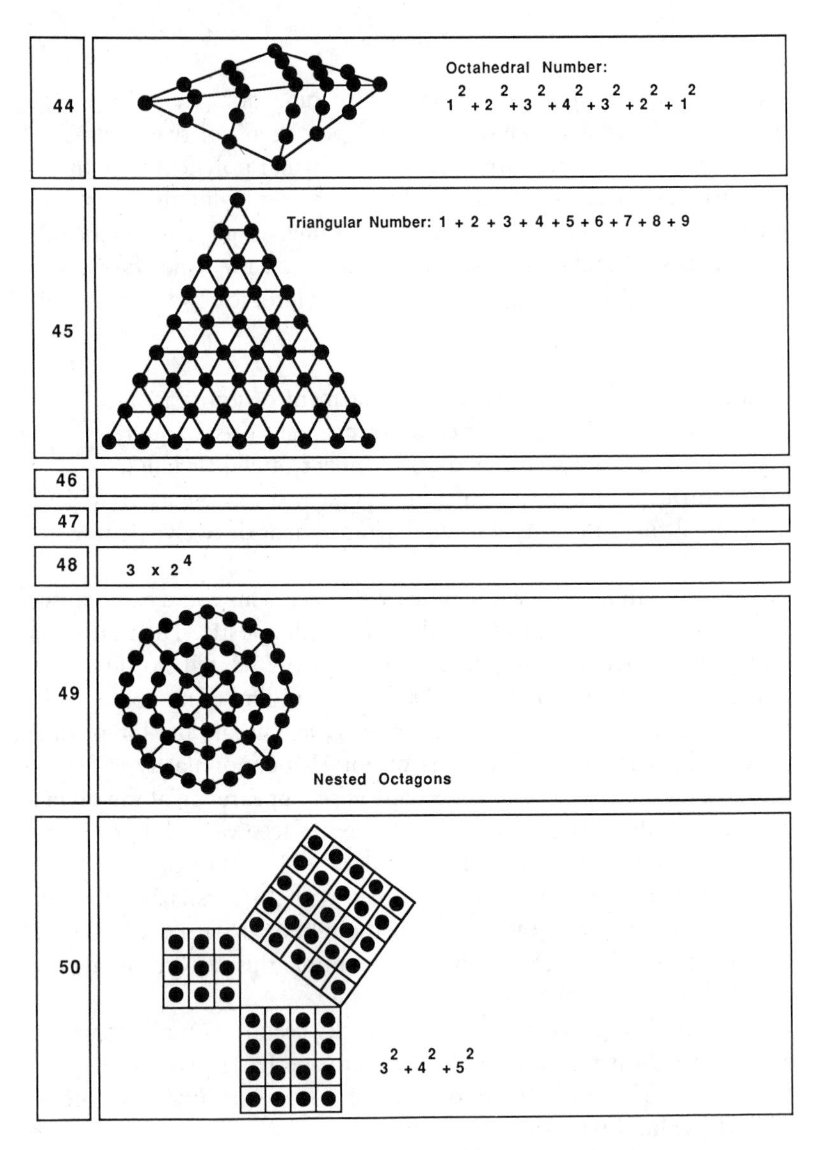

Fig. 29, No. 10.

cornered pattern that represents a $2 \times 2 \times 2 \times 2$ hypercube. Sixteen can also be represented by a pattern of nested pentagons. In our society 16 is important as the age when a person can get a driver's license,

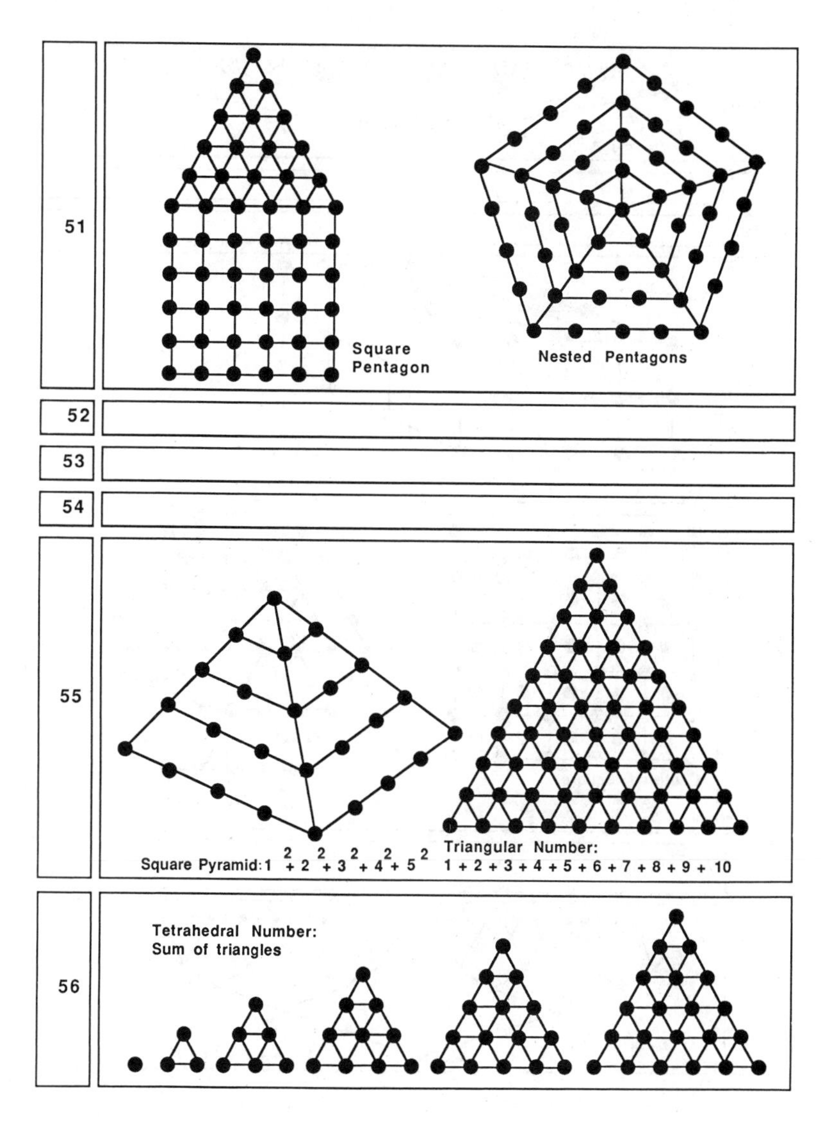

Fig. 29, No. 11.

and "Sweet Sixteen" is often thought of as the age when a girl reaches womanhood.

There is really very little to say about 17. It is 8 plus 9, which is somewhat interesting, I suppose. Alternatively, one might represent

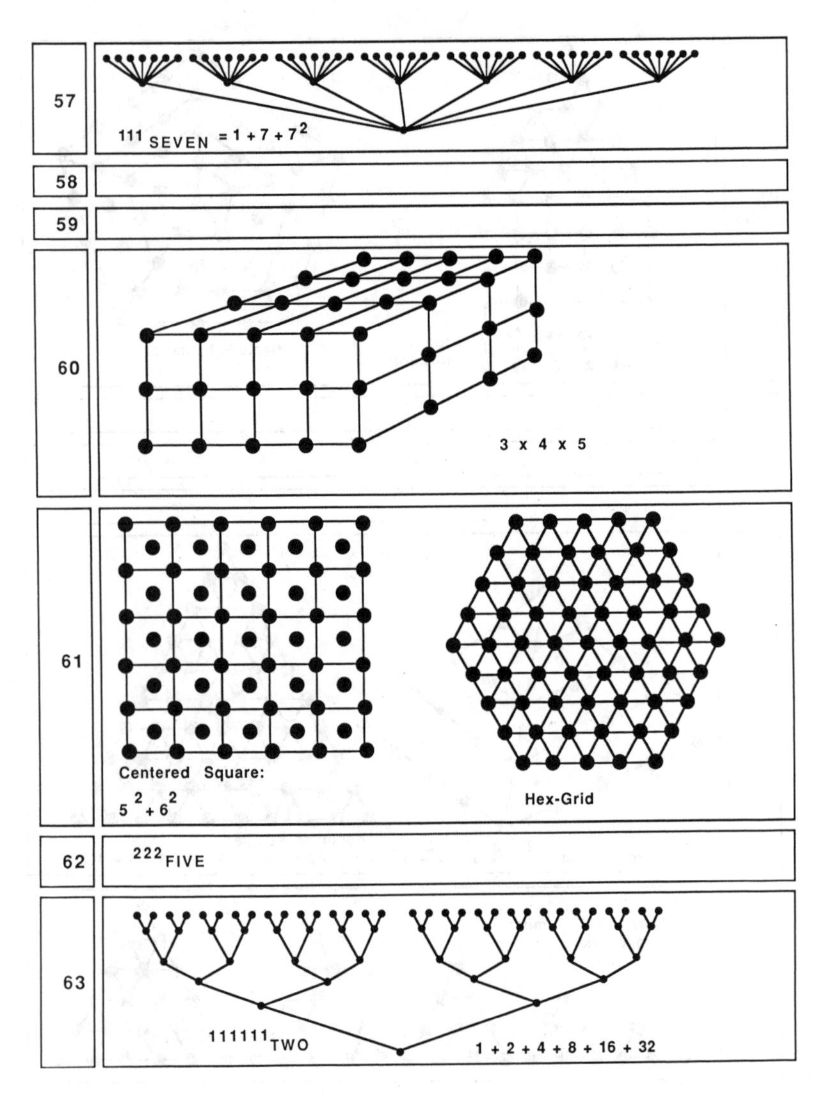

Fig. 29, No. 12.

17 as a "casket" pattern: two 2 × 2 squares on either side of a 3 × 3 square.

Eighteen also provides slim pickings, but 19 is interesting in two ways. First, 19 dots fit into a nice hexagonal grid pattern; second, 19 dots can be arranged into an octahedral pattern.

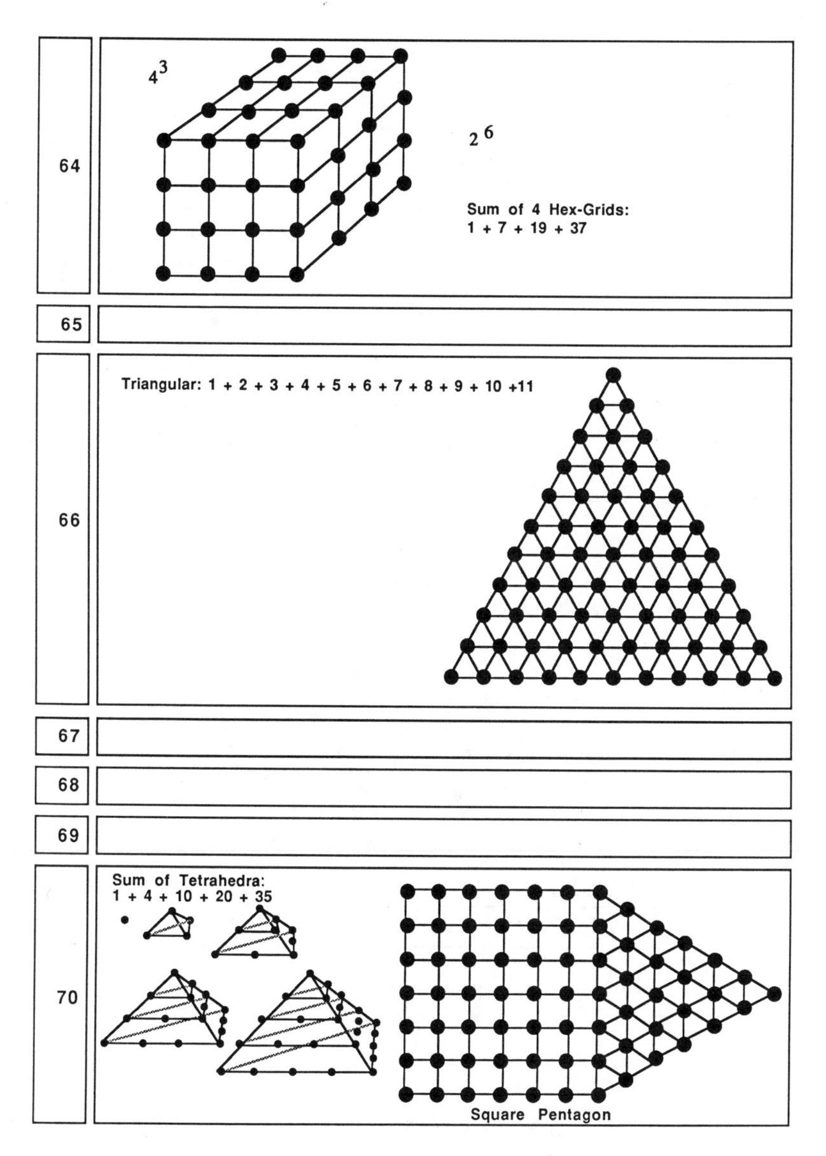

Fig. 29, No. 13.

Twenty is significant as the number of corners on a dodecahedron; another thing about 20 is that it is a "tetrahedral" number. We say that a number is tetrahedral if it is the sum of successive triangular

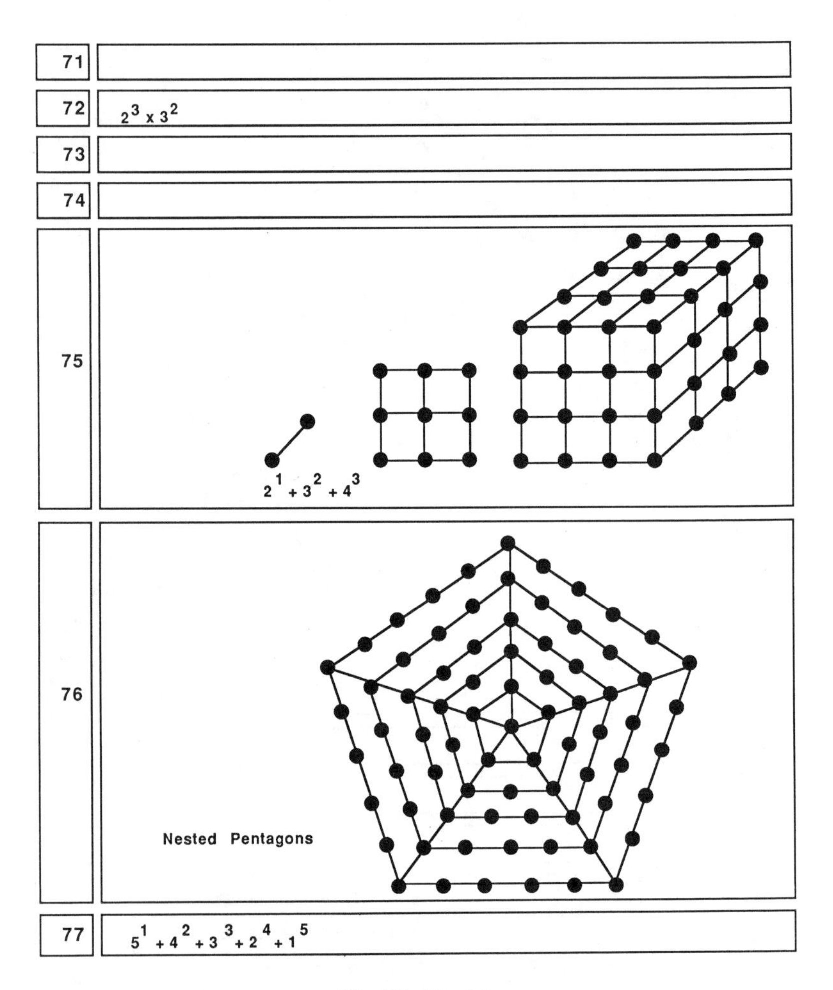

Fig. 29, No. 14.

numbers. Twenty spheres (try tennis balls or oranges) can be stacked in a tetrahedral pattern.

As we move further down our list of pictures, we find numbers that embody no simple pattern at all. The first really hard case is 23. For some reason 91 is one of the most exciting numbers of all!

It is interesting, even mesmerizing, to see how many patterns and connections are hidden in our first 100 numbers. I've thought about

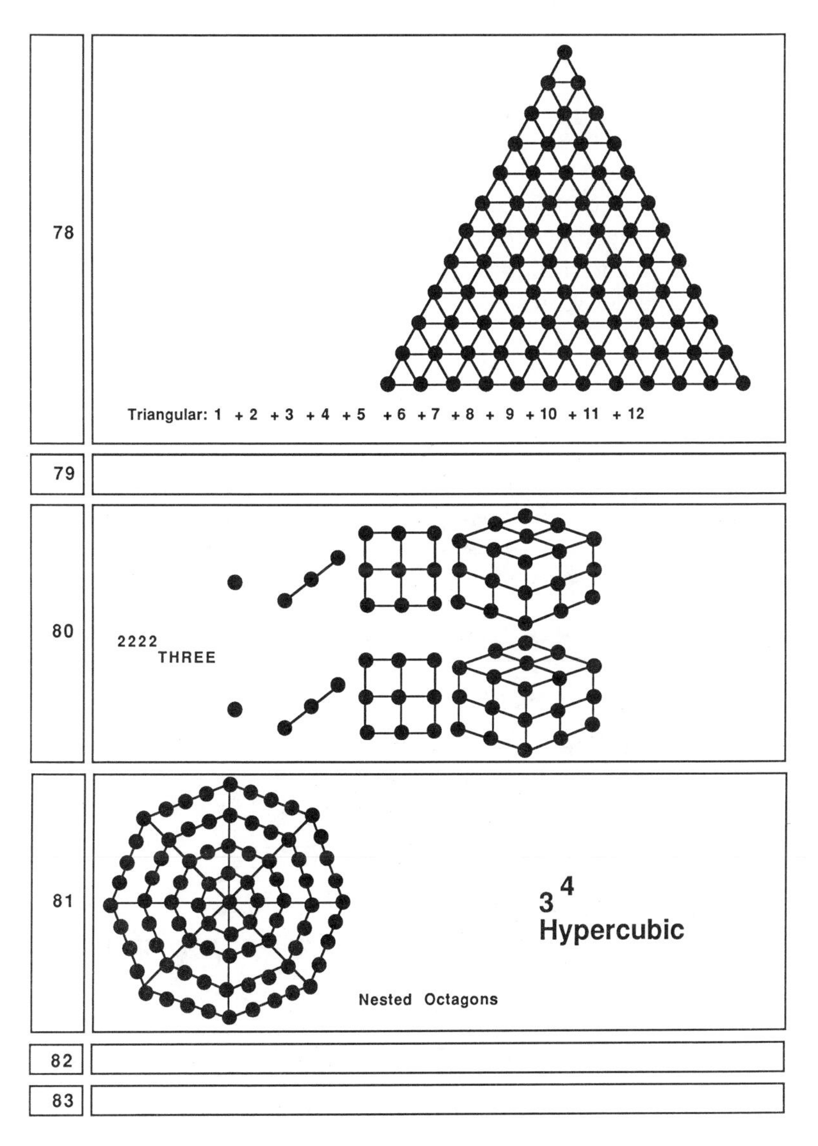

Fig. 29, No. 15.

nothing else for a couple of weeks now, and I'd like to stop. Last night I got so bothered about 11's lack of a good pattern that I got up out of bed to check that a tetrahedron of four billiard balls fits nicely on top of a hexagonal grid of seven billiard balls, which is something, I suppose.

But what good is all this? A good, if superficial, use for these patterns is in decorating birthday and anniversary cards. People enjoy knowing that their age (or number of years married) embodies some interesting geometrical form.

At a deeper level, this exercise gives one a much better insight into

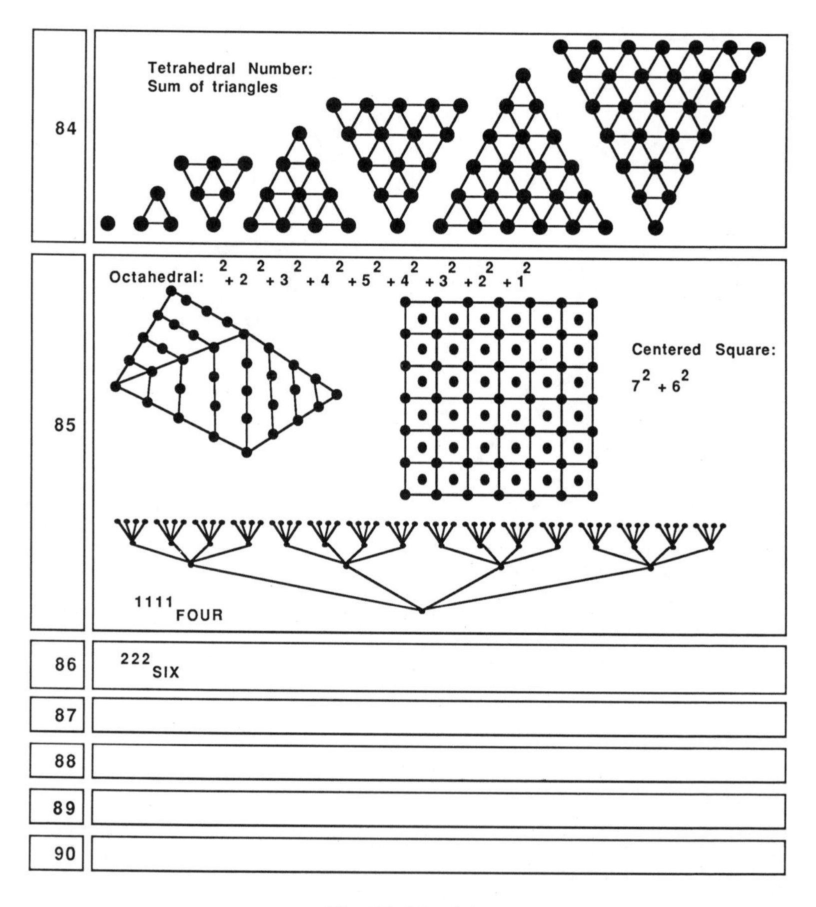

Fig. 29, No. 16.

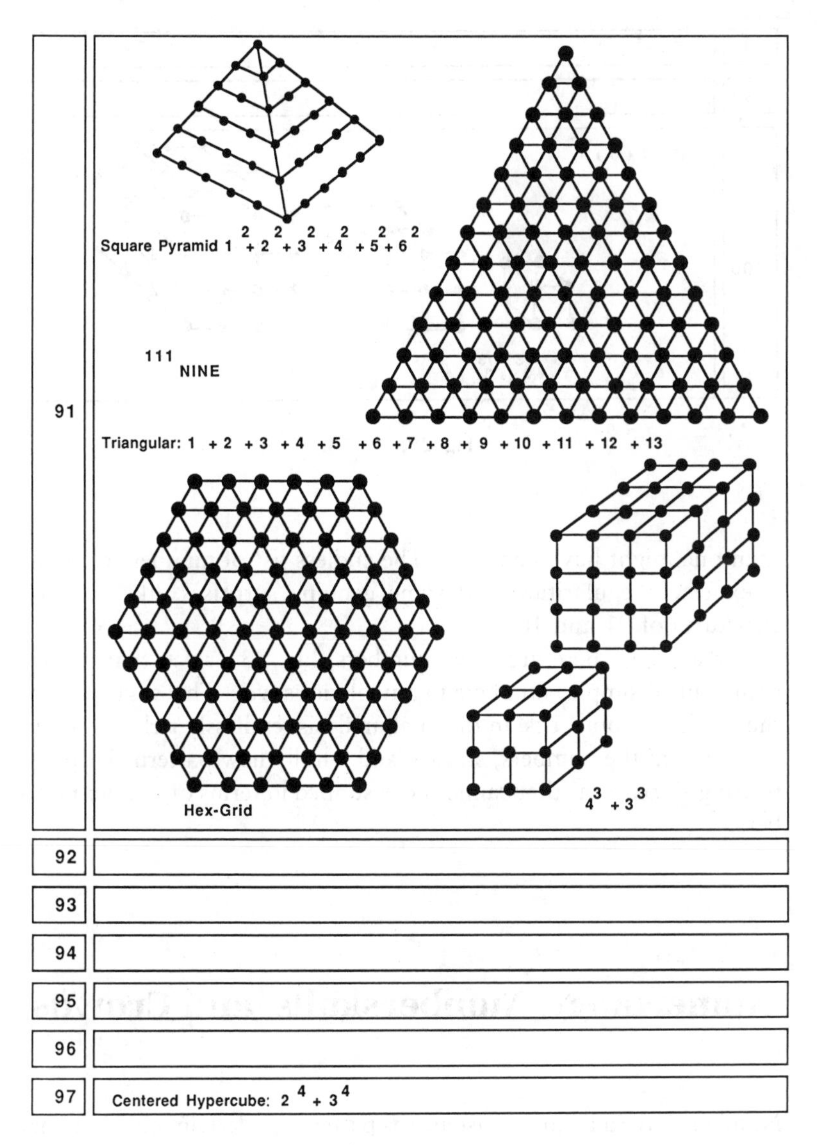

Fig. 29, No. 17.

the nature of numbers. All the patterns I drew are based on combinations of small numbers. Small numbers are added, multiplied, squared, and cubed. These simple combinations yield many more patterns

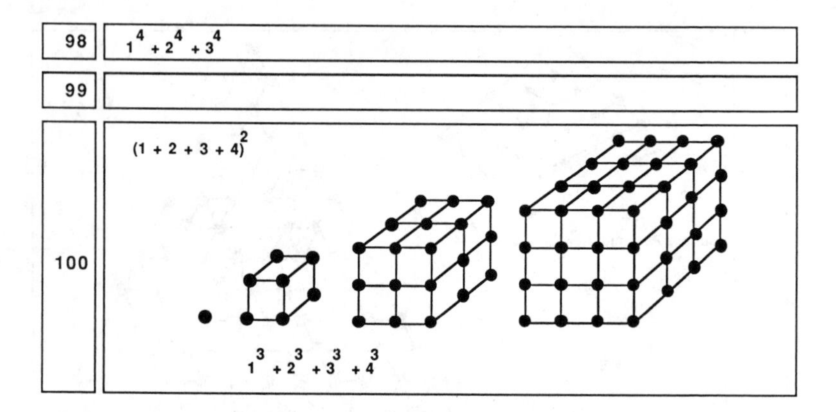

Fig. 29, No. 18.

than one might have imagined. The endless list of numbers contains unexpected regularities, and some genuine surprises, like the rich structures of 91 and 100.

For all this, there are some numbers, like 23 or 59, that do not come out of simpler numbers in any obvious way. This suggests that the world of number is, in the final analysis, endlessly rich. As we go out through the numbers, strange and wholly new patterns keep appearing — patterns that cannot be described in terms of smaller numbers.

Numerology, Numberskulls, and Crowds

Numbers have a definite existence as patterns underlying the thoughts and objects that surround us. Depending on the complexity of the pattern from which they are drawn, numbers fall into four rough size ranges: small, medium, large, and inconceivable. *Small numbers* — the numbers, let us say, from 1 through 1000 — code up simple, universal archetypes. These numbers have the solidity and definiteness of naturally occurring crystals. *Medium numbers* — the numbers from

1000 up through a trillion or so — are used to stand for various collections of discrete objects: the number of trees in a forest, the number of people on earth, the number of books in a library, the gross national product in dollars, and so on. These numbers are hard to visualize "all at once," and they have a more abstract quality than do the small numbers. *Large numbers* are the numbers lying out at the fringes of our ability to come up with number names. "Googol" is one of the smallest large numbers. Googol is written as a 1 followed by a hundred 0s: 10,000,000,000,000,000,000,000,000,000,000, 000,000,000,000,000,000,000,000,000,000,000,000,000,000,000, 000,000,000,000,000,000,000,000. These large numbers, as we will see in "Words as Numbers," are important not so much for their brute size as for the patterns they embody. Viewed as a pattern of information, a book is really a special kind of large number, some million "digits" long. *Inconceivable numbers* are the numbers that, while not infinite, are nevertheless so large and complex that we have no clear way of thinking about them. I won't say anything else about them in this section.

Academic philosophers sometimes write articles arguing that numbers are fictional thought forms contingent on the human race for their very existence. This seems false and a bit chauvinistic. No one who really knows anything about mathematics can think that numbers are things we "just make up." If you play around with numbers for just a little while, you begin to realize how stubbornly objective their properties are. Three dots will not form into a square pattern, no matter how much any human may want them to. Only a landlubber can disbelieve in whales, and only a tenured philosophy professor can disbelieve in numbers!

Other people go to an opposite extreme and argue that everything *except* numbers is an illusion. The followers of Pythagoras seem to have held a view something like this. They thought of numbers as tiny dot patterns and regarded the world as built up from the various dot patterns like a toy castle made out of Lego blocks. This world view was to some extent vindicated by the discovery that matter is built up out of atoms of the various chemical elements, and by the discovery that the elements can be arranged so that each one has a well-defined atomic *number* — oxygen has atomic number 16, gold has atomic number 79, and so on. Of course, when we go down to the subatomic level, things are no longer so crystalline and Pythagorean. Still, it is pretty remarkable that so much of chemistry can be explained in terms of patterns of small numbers.

Superstitious people go much further than this and try to explain everything in the world by manipulations of small numbers. This mode of thinking is known as "numerology." Here a person begins to think that things like dates, street addresses, and Bible-verse numbers have very great significance. The basic notion is that the world is a magical pattern of small numbers arranged in simple patterns. Gamblers often fall prey to this obsession, particularly those who bet on state-sponsored lotteries or on the illegal, privately run lotteries known as "the numbers."

Most cities of any size have "numbers" games. To bet in such a game, a person picks a three-digit number and gives an agent a small amount of money. The winning number is taken from a public source — sometimes the last three digits of the daily federal treasury balance is used. Although the odds of winning are 1000 to 1, the payoff is usually at a rate of 500 to 1.

Despite the poor odds, a lot of people play the numbers. It is, in point of fact, the most popular form of gambling in America. Numbers players, like other gamblers, often get the feeling that by tuning in to the right cosmic rhythms they can beat the system. Entrepreneurs capitalize on this feeling by selling "Numbers Books" on city newsstands. Books like this consist, basically, of long lists of words with a three-digit number assigned to each word. Thus, if you dream of a dog, you might look up "dog" and find the number 328, or if you see a hearse passing in the street, you might look up "hearse," and find the number 024.

Much more complicated kinds of numerological notions are involved in traditional occult systems like the cabala and the tarot, but I don't know enough about these systems to make any interesting observations on them. Instead, I would like to talk next about a number obsession somewhat different from numerology. Numerology deals with small numbers; the obsession I am now thinking of deals with medium numbers. This obsession is very common, but there is no word for its victims. I propose that we call them *numberskulls.*

As an example of a numberskull, think of a cousin who, when he comes to see you, talks about the route numbers of the roads he took and about his car's gas mileage, about the won–lost record of his favorite team, about the fine points of his business's tax rates, about the number of Chinese being born every day, about the number of miles per week he's been jogging, about the price per case of the wine he brought you, about his daughter's grade-point average, about how many copies the record you play has sold, about the value of the mortgage still outstanding on your home, about the federal deficit, and so on — numbers, numbers; medium numbers; numberskull numbers all evening long!

Most of our military and political leaders appear to be numberskulls. It comes naturally to them. Troops or weapons lined up on a parade ground form a vast grid of identical units. To force someone to vote is to reduce his or her complex opinions to a single numbered

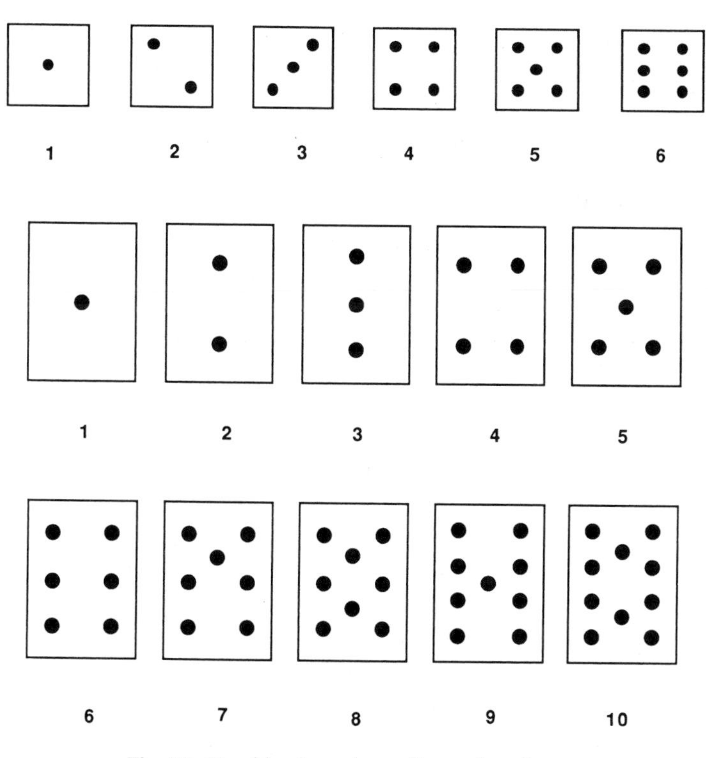

Fig. 30 Gamblers' numbers: dice and cards.

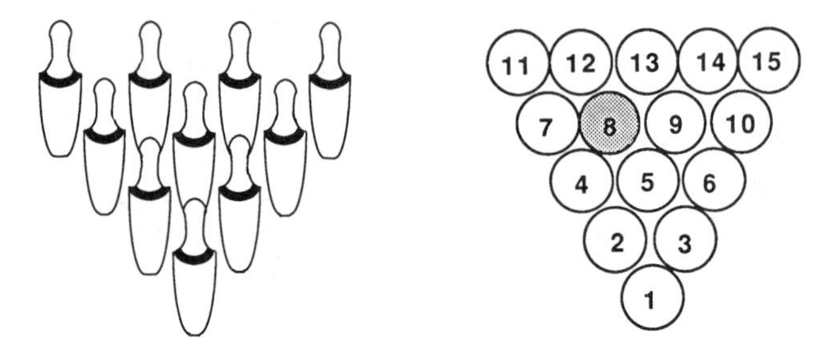

Fig. 31 Sportsmen's numbers: bowling and pool.

choice. Every newspaper article bristles with numbers: dollars, percentages, rates of growth, megatons.

It is hard to imagine what the world would be like if it weren't run by numberskulls. Surely, to take the most obvious example, it is the sheer hypnosis of numbers that drives the arms race ever onward. Who but a numberskull could care more about megatons than about radiation sickness? Here's a numberskull fact: 1 nuclear megaton = 2,000,000,000 pounds of TNT. One-half pound of TNT set off near a person is lethal; therefore one megaton is enough to kill four billion people.

The millions and billions that the numberskull loves so well are dull and lifeless things. The numberskull's numbers do not have the transparent structure of the small numbers, nor do they have the subtle patterning of the really large numbers. The medium numbers are big, messy heaps with short, boring names. Thinking in terms of these numbers gives one an exaggerated estimate of one's own powers. The numberskull falls into the illusion that his talk about millions is puissant enough to effect real-world changes. This is a false and dangerous dream.

When experienced in a direct, tactile way, medium numbers are something quite different from the figures one reads in the newspapers. There is a strangely energizing pleasure in joining a crowd of 10,000 people. Being in a large crowd of like-minded people is a good part of the pleasure one gets from attending a concert, a sporting event, or a religious ceremony. In India there is a special word for this feeling: *darshan*, meaning the mystical experience of being in a very large gathering. As Elias Canetti puts it, in his classic study *Crowds and Power:* "A man attending a sermon honestly believed that it was the sermon which mattered to him, and he would have felt astonished or even indignant had it been explained to him that the large number of listeners present gave him more satisfaction than the sermon itself. All ceremonies and rules pertaining to such institutions are basically intent on capturing the crowd."

Canetti distinguishes between "open" and "closed" crowds. A closed crowd, such as a church congregation, is limited in size. An open crowd, such as a street demonstration, is characterized by its desire to grow: "As soon as it exists at all, it wants to consist of *more* people: the urge to grow is the first and supreme attribute of the crowd. It wants to seize everyone within reach; anything shaped like a human being can join it. The natural crowd is the *open* crowd; there are no limits whatever to its growth; it does not recognize houses, doors or locks and those who shut themselves in are suspect. . . . The open crowd exists so long as it grows; it disintegrates as soon as it stops growing." Numbers have this same quality of wanting to grow. Once you start counting, it's hard to stop.

The Mind Reckoner

Children often compete with each other to see who can name the biggest number. It's a game that some of us never quite outgrow.

At a crude, magical level, knowing names for big numbers seems to give one a kind of power over the world. Maybe I can control the heavens if I know how many stars are in our galaxy (about 100 billion)? Perhaps I could win more often at bridge by knowing the number of all possible hands (a bit under one quadrillion)? Maybe I could run faster if I knew how many atoms were in one of my breaths (about one sextillion)? Perhaps I could give up drinking if I knew how many molecules were in a can of beer (about ten septillion)?

One of the earliest efforts at naming large numbers occurs in *The Sand Reckoner,* a treatise Archimedes wrote for his friend Zeuxippus in 215 B.C. so as to favorably impress the king Gelon. Here Archimedes reckoned that the universe (then believed to be about one light-year across) has room for some vigintillion grains of sand, a vigintillion being 1,000,000,000,000,000,000,000,000,000,000,000,000, 000,000,000,000,000,000,000,000,000, one followed by sixty-three zeros. This number lies at the border of what I have called medium and large numbers. Archimedes was the first person to get this far. His pride in his accomplishment is evident in his treatise's opening passage:

> There are some, King Gelon, who think that the number of the sand is infinite in multitude; and I mean by the sand not only that which exists about Syracuse and the rest of Sicily but also that which is found in every region whether inhabited or uninhabited. Again there are some who, without regarding it as infinite, yet think that no number has been named which is great enough to exceed its multitude. And it is clear that they who hold this view, if they imagined a mass made up of sand in other respects as large as the mass of the earth, including in it all the seas and the hollows of the earth filled up to a height equal to that of the highest of the mountains, would be many times further still from recognising that any number could be expressed which exceeded the multitude of the sand so taken. But I will try to show you by means of geometrical proofs, which you will be able to follow, that, of the numbers named by me and given in the work which I sent to Zeuxippus, some exceed not only the number of the mass of sand equal in magnitude to the earth filled up in the way described, but also that of a mass equal in magnitude to the universe. (This translation appears in J. Newman, *The World of Mathematics*.)

The Sand Reckoner was the last thing Archimedes wrote — a few years later some Roman soldiers killed him while he was drawing figures in a tray of sand. One of the interesting things about *The Sand Reckoner* is that Archimedes not only had to calculate how many grains of sand would fit into a sphere the size of the Greek's "universe," he also had to coin some new number names to describe the number he arrived at.

In this section I'd like to imitate Archimedes by assigning a number to a very large, seemingly infinite collection: the collection of all humanly possible thoughts. To assign a number to this collection, it will be necessary to develop some new number names. Let's begin by going over the large-number names we already have.

1	One
10	Ten
100	Hundred
1,000	Thousand
1,000,000	Million
1,000,000,000	Billion
1,000,000,000,000	Trillion
1,000,000,000,000,000	Quadrillion
1,000,000,000,000,000,000,000	Sextillion
1,000,000,000,000,000,000,000,000	Septillion
1,000,000,000,000,000,000,000,000,000,000,000	Decillion
1,000,000,000,000,000,000,000,000,000,000,000, 000,000,000,000,000,000,000,000,000,000	Vigintillion
1,000,000,000,000,000,000,000,000,000,000,000, 000,000,000,000,000,000,000,000,000,000, 000,000,000,000,000,000,000,000,000,000, 000,000,000,000,000,000,000,000,000,000, 000,000,000,000,000,000,000,000,000,000, 000,000,000,000,000,000,000,000,000,000, 000,000,000,000,000,000,000,000,000,000, 000,000,000,000,000,000,000,000,000,000, 000,000,000,000,000,000,000,000,000,000, 000,000,000,000,000,000,000,000,000,000	Centillion

The principle behind the "-illion" number names is that, for any number prefix *N*, an "*N*-illion" stands for the number written as one with $3N + 3$ zeros. For $N = 1, 2, 3, 4, 6, 7, 10, 20$, and 100, the number prefixes are, respectively, mi-, bi-, tri-, quadri-, sex-, sept-, dec-, vigint-, and cent-; these "-illions" have, respectively, 6, 9, 12, 15, 21, 24, 33, 63, and 303 zeros.

People have occasionally tried pushing the "-illion" system out to very great lengths. One difficulty in doing this is that it is hard to decide on exactly what number prefixes one should use for the larger numbers; another difficulty is that, instead of just having *N* zeros, *N*-illion has to have $3N + 3$ zeros. This is fine at the lower levels, but we would like to have some simple name for things like one followed by a hundred zeros, or one followed by a thousand zeros.

As was mentioned in the last section, the number written as one followed by a hundred zeros is known as a *googol:* The mathematician Edward Kasner popularized this number name in the 1930s. Googol is a very big number — most scientists agree that if we could count up all the atoms in all the stars we can see, we would come up with less than googol of them. If we really want to, we can handle googol in terms of the -illion system by noting that $100 = 1 + 3(32) + 3$, so that googol is really ten 32-illion. Instead of using "32" for the number prefix, we could use the Latinate "triginti-duo" and call googol the same as ten triginti-duillion, but this is very awkward. Kasner and Newman's book, *Mathematics and the Imagination,* suggests a way out: "Words of wisdom are spoken by children at least as often as by scientists. The name 'googol' was invented by a child (Dr. Kasner's nine-year-old nephew) who was asked to think up a name for a very big number, namely, 1 with a hundred zeros after it. He was very certain that this number was not infinite, and therefore equally certain that it had to have a name. At the same time that he suggested 'googol' he gave a name for a still larger number: 'googolplex.' The googolplex is a specific finite number, with so many zeros after the 1 that the number of zeros is a googol."

I propose that we go ahead and use the -plex notation as freely as the -illion notation. The basic idea is simple. If *N* is a number prefix, we will let *N*-plex be the number written as one followed by *N* zeros. In -plex notation, ten, hundred, thousand, million, and billion are, respectively, oneplex, twoplex, threeplex, sixplex, and nineplex, as they have, respectively, 1, 2, 3, 6, and 9 zeros. Googol is the same as hundredplex.

Put a bit differently, *N*-plex is simply 10^N, and 10^N, pronounced "ten to the *N*th power," stands for the product of *N* tens, which is the number written as one with *N* zeros. There are two good reasons for using *N*-plex instead of 10^N: *N*-plex can be written without superscripts, and *N*-plex is easy to say.

The -illion numbers have prefixes based on Latin number names. Usually, we'll just use ordinary number names for the -plex prefixes, but it will be fun to use special number prefixes for a few of them. The special prefixes I have in mind are based on Greek number names; they were adopted by an international committee on weights and measures in 1958. The standard Greek prefixes for ten, hundred, thousand, million, billion, and trillion are, respectively, deka-, hecto-, kilo-, mega-, giga-, and tera-, so we will talk about dekaplex, hectoplex, kiloplex, megaplex, gigaplex, and teraplex.

10^{10}	dekaplex, also known as ten billion
10^{100}	hectoplex, also known as googol
10^{1000}	kiloplex
$10^{1,000,000}$	megaplex
$10^{1,000,000,000}$	gigaplex
$10^{1,000,000,000,000}$	teraplex
10^{googol}	googolplex, also known as hectoplexplex

A nice thing about logarithms is that they make it easier to talk about large numbers. If we are taking our logs in base ten, then log *N*-plex is *N*. If we repeat the log operation, then further reductions are possible. Thus, log googol is 100, so log (log googol) is log (100), which is 2. Pushing it another step, we see that

$$\begin{aligned}\text{log log log googolplex} &= \text{log log googol}\\ &= \text{log } 100\\ &= 2.\end{aligned}$$

If we use a number line whose scale shifts from regular to logarithmic to logarithmic of logarithmic, and so on, we can fit most of our large numbers into a compact illustration, such as Fig. 32.

Now I would like to estimate how many different possible thoughts a person might have. Let's take a very crude approach. The brain has, it is said, about three billion synapses. Suppose that a thought, or state of mind, can be characterized by stating exactly which synapses are firing. At this very crude level, we might think of each of the three billion synapses as being like an on-off switch in a computer's central processing unit. To specify a state of mind we say whether the first synapse is on or off, whether the second synapse is on or off, and so on. This makes for two to the three-billionth possibilities, which is a number in the gigaplex range. We might say, then, that there are around gigaplex possible states of mind:

$$2^{(3\text{ billion})} = (2^3)^{\text{billion}} \approx 10^{\text{billion}} = \text{gigaplex}.$$

What, again, is a gigaplex? A gigaplex is the number written as one with a billion zeros. It feels as if a person might have an endless number of different thoughts, but unless one is somehow to enlarge the brain, there are really only gigaplex possible states of mind.

In 1984 there was a science-fiction movie called *Brainstorm,* which was Natalie Wood's last film. The movie centered around the invention of a brainstate recorder. By taping a person's brain activity and playing it back, you could experience the person's life from his or her own point of view. Looked at in a certain way, you might say that a person's life could be fully coded up by such a recording, for playing the recording would feel exactly the same as living the life in question.

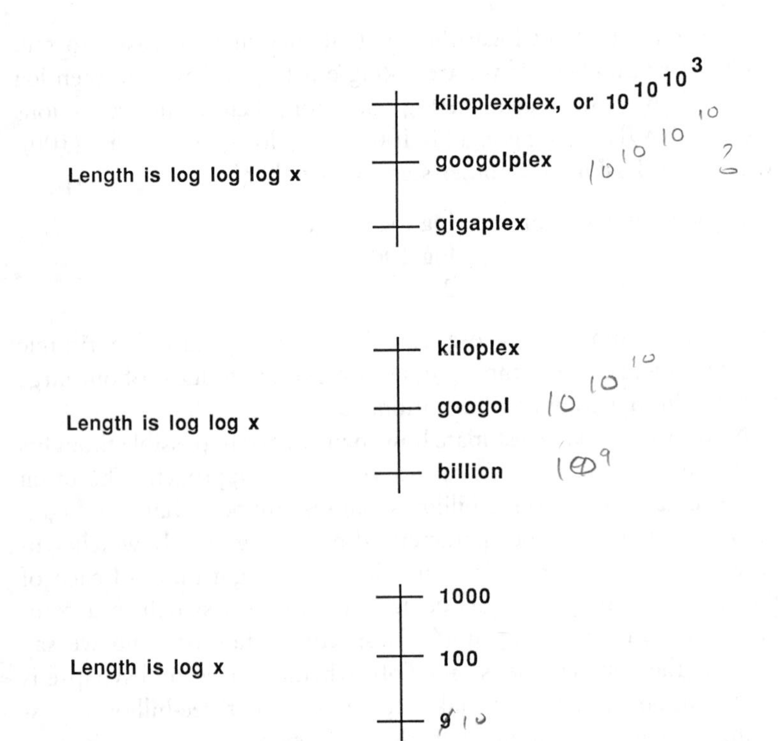

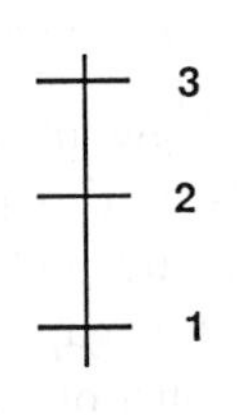

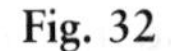

Fig. 32

Our brainstates do not change all that rapidly or unpredictably, so it would probably be enough to record the brainstate, say five hundred times a second. A billion seconds lasts just under thirty-two years, so a thousand billion brainstate samples would pretty well take care of an average person's life. That is to say, a trillion samples would fully specify a life.

If we go with our prior estimate that there are gigaplex possible choices of brainstate for each sample instant, we come up with gigaplex to the trillionth power possible human lives, which is sextillionplex:

$$\begin{aligned}(\text{gigaplex})^{\text{trillion}} &= (10^{1{,}000{,}000{,}000})^{1{,}000{,}000{,}000{,}000} \\ &= 10^{(1{,}000{,}000{,}000 \times 1{,}000{,}000{,}000{,}000)} \\ &= 10^{1{,}000{,}000{,}000{,}000{,}000{,}000{,}000} \\ &= 10^{\text{sextillion}} = \text{sextillionplex.}\end{aligned}$$

It's kind of neat to have a number for things like this. For a human, there are gigaplex possible thoughts, and sextillionplex possible lives. That's really worth knowing. It's a striking result, even though the math is simple. As Archimedes concluded, "I conceive that these things, King Gelon, will appear incredible to the great majority of people who have not studied mathematics, but that to those who are conversant therewith . . . the proof will carry conviction."

Words as Numbers

Let's start this section with an old joke.

A man is sent to prison for a ten-year term. The dining hall also serves as an auditorium, and there is a stage at one end. After supper, one of the prisoners runs up onto the stage and hollers, "Four hundred and eighty-seven." Everyone starts laughing. The next day, it's the same thing: After supper someone jumps onto the stage, yells, "Two thousand six hundred and twenty-two," and all the prisoners crack up. This goes on for a couple of weeks, and finally the man asks his cellmate what's going on.

"Well," says the cellmate, "It's like this. The prison library has a big fat book called *The Ten Thousand Best Jokes,* and we've all been here so long that we know the book by heart. If somebody wants to tell a joke, they just shout out the number of the joke in *The Ten Thousand Best Jokes,* and if it's a funny one, everybody laughs."

At dinner that night, the man decides to show the other prisoners that he's a good guy. Before anyone else can get to the stage, he dashes up there and shouts, "Five thousand nine hundred and eighty-six!" But to his horror, nobody cracks a smile. There are even a few groans. He slinks back to his cell to consult with his cellmate.

"Nobody laughed! Isn't the five thousand nine hundred and eighty-sixth joke a good one?"

"Sure it's a good one," says the cellmate. "Old five thousand nine hundred eighty-six is one of the best."

"So why didn't anyone laugh?"

"You didn't tell it right."

The difference between "knowing the number" and "telling it right" is another example of the digital–analog split we talked about in the Introduction, but the real relevance of this story here is that, given a coding system, numbers can be used to code up all kinds of things. In the story above, the coding system involved looking things up in a special code book. The familiar system of rating things on a scale from 1 to 10 involves a code book that is not explicitly given, but might run something like this:

1 The worst
2 Very bad
3 Bad
4 Poor
5 So-so
6 OK
7 Decent
8 Good
9 Really good
10 The best

So far as I know, in recent times the 1-to-10 rating system had its greatest vogue in conversations between US soldiers and South Vietnamese peddlers. Even if people have no language at all in common, they know the same numbers.

If two people have copies of the same dictionary, then they can exchange messages by giving the position numbers of the words they want to use. Thus, using my Merriam–Webster collegiate dictionary, "how are you" might become "30,068–3480–74,462." This is, of course, a very cumbersome way of coding; it's hard to always carry around a fat dictionary for a code book.

Just writing words out is really a kind of numerical code in itself. There is no great distinction between numeral symbols and letter symbols — they are all simple squiggles that are easy to draw by hand and easy to tell apart. Relative to a numeration system based on 27, the blank space and the letters of the alphabet could be thought of as individual numerals, and pieces of text could be thought of as base-twenty-seven numbers. To simplify, let us suppose that we will write our text in lower-case letters, with "–" marks to stand for spaces. We could think of "–" as 0, "a" as 1, "k" as 11, and so on. There is, abstractly speaking, nothing wrong with basing a numeration system on a base greater than ten, and there is also nothing wrong with using the letter squiggles to stand for numerals. Suppose that we call this special numeration system the "L27 system," where the "L" means we use letter symbols for numerals, and the "27" means that the system will be based on powers of 27. As we will see shortly, in the L27 system a string of letters like "big" stands for a definite number (1,708, in the case of "big").

To understand this odd numeration system, keep in mind the analogy to our familiar base-ten system. In the L27 system, "–" and "a" have the same force as do 0 and 1 in base ten, and "z," which is one less than the number base (twenty-seven), has the same force as does 9 in the base-ten system. Thus, passing from "z" to "a–" is analogous to the passage from 9 to 10, and going from "zzz" to "a–––" is like going from 999 to 1000. The whole difference in L27 is, again, that we are using letter symbols as numerals, and we are working with powers of 27 (1, 27, 729, 19683, 531441, . . .) rather than with powers of 10 (1, 10, 100, 1000, 10,000, . . .). Recall from our discussion of logarithms that a decimal digit carries $\log_2 10$, or 3.32 bits of information. An L27 digit carries $\log_2 27$, or 4.75 bits of information. Since there is more information per symbol, a short list of L27 symbols can name a very large number.

As a matter of terminology, what should we call a list of L27

Table 1. The L27 Numeration System

L27 symbol	*Base-ten symbol*	*L27 symbol* (cont.)	*Base-ten symbol* (cont.)
– (blank)	0	a–	$1 \times (27) = 27$
a	1	aa	$1 \times (27) + 1 = 28$
b	2	ab	$1 \times (27) + 2 = 29$
c	3	...	...
d	4	az	$1 \times (27) + 26 = 53$
e	5	b–	$2 \times (27) = 54$
f	6	ba	$2 \times (27) + 1 = 55$
g	7	...	...
h	8	hi	$8 \times (27) + 9 = 225$
i	9	...	...
j	10	z–	$26 \times (27) = 702$
k	11	za	$26 \times (27) + 1 = 703$
l	12	...	...
m	13	zz	$26 \times (27) + 26 = 728$
n	14	a––	$1 \times (27 \times 27) = 729$
o	15	a–a	$1 \times (27 \times 27) + 1 = 730$
p	16	...	...
q	17	aa–	$1 \times (27 \times 27) + 1 \times (27) = 756$
r	18	aaa	$1 \times (27 \times 27) + 1 \times (27) + 1 = 757$
s	19	...	...
t	20	are	$1 \times (27 \times 27) + 18 \times (27) + 5 = 1220$
u	21	...	...
v	22	big	$2 \times (27 \times 27) + 9 \times (27) + 7 = 1708$
w	23	...	...
x	24	zap	$26 \times (27 \times 27) + 1 \times (27) + 16 = 18997$
y	25	...	...
z	26	zzz	$26 \times (27 \times 27) + 26 \times (27) + 26 = 19682$
		a–––	$1 \times (27 \times 27 \times 27) = 19683$

symbols? It seems a bit unnatural to call it a *number,* so suppose we call it a *string.*

If you have a calculator to hand, you might figure out the base-ten number your name stands for, if you write it lower-case and view it as an L27 string. My name is 369,736:

$$\begin{aligned} \text{rudy}_{L27} &= 18 \times (19683) + 21 \times (729) + 4 \times (27) + 25 \\ &= 369{,}736. \end{aligned}$$

Pleased to meet you.

It is also possible to reverse this coding process. Given any base-ten number, a little trial-and-error calculation will reveal the unique L27 string associated with it. If, for instance, you know the L27 system, and I want to say "hi" to you, I could hand you a box with 225 marbles in it. Before too long, you would realize that the 225 marbles break into 8 groups of 27 marbles, with 9 single marbles left over, and this means that, relative to the L27 system, I have given you "hi" marbles. If you like to imagine that the world is filled with secret clues, then you might enjoy turning various number collections into L27 strings:

$$\begin{aligned} 225 &= 8 \times (27) + 9 = \text{hi}_{L27}; \\ 1987 &= 2 \times (729) + 19 \times (27) + 16 = \text{bsp}_{L27}; \\ 666 &= 24 \times (27) + 18 = \text{xr}_{L27}. \end{aligned}$$

So "xr" is the L27 Number of the Beast! I'm just glad my first name isn't "Xavier."

A different way of looking at the L27 counting system is that it systematically lists all possible combinations of the blank and the twenty-six letters, in a kind of dictionary order. I say a *kind* of dictionary order because this listing works by first listing all the strings that are one symbol long, then all the strings that are two symbols long, and so on.

It will be useful to think of any piece of text as being an L27 string. One small problem is that the L27 system does not have punctuation or any other special symbols, nor does it have capital letters. Really, though, as long as we have spaces, we can live without capitals and punctuation, and any really important special symbol can always be replaced by its name. We can put "equals" instead of "=," or "five" instead of "5."

To treat an arbitrary piece of text like an L27 string, then, we will write all capitals in lower-case, replace all punctuation marks with spaces, and replace any special symbols (such as numerals) with their names. By this method, a heavily punctuated passage such as ended the third paragraph before this one becomes "so––xr––is–the–ltwoseven–number–of–the–beast–––i–m–just–glad–my–first–name–isn–t––xavier––"

This is clearly an L27 string, consisting only of spaces and lower-case letters. It is 90 symbols long, so it stands for a number that starts with a multiple of 27 to the 89th power:

$$19 \times (27^{89}) + 15 \times (27^{88}) + \cdots + 18 \times (27^{2}) + 0 \times (27) + 0 \approx 27^{90} \approx 10^{129}.$$

This number, which we might call the L27 number value of the passage in question, is very large, bigger than googol, though still less than kiloplex.

Earlier I mentioned that I might say an L27 "hi" by handing you 225 marbles, but I certainly would not want to have to hand over more than googol marbles just to make a trivial joke about my name and the number 666. Why *is* the L27 number value for a short message so very large? The reason is that the L27 number value is approximately equal to the total number of possible messages of the same length that might have been sent. There are 27 to the 90th power possible ways to write an L27 list 90 symbols long. When you write out a piece of text, you are really creating an L27 string. You are, whether you realize it or not, choosing one special message out of an astronomical number of possibilities.

"Astronomical" is really too weak a word here. A short paperback novel of seventy thousand words is some seven hundred thousand symbols long, and this makes for an L27 number value about equal to megaplex. There are, in other words, about megaplex possible short paperbacks, where any possible string of L27 symbols at all is being allowed:

$$27^{700,000} \approx (10^{1.43})^{700,000} = 10^{(1.43 \times 700,000)} \approx 10^{1,000,000} = \text{megaplex}.$$

Recall here that in "The Mind Reckoner" we estimated that there are gigaplex possible human thoughts. Relating that estimate to the type of estimate just made indicates that a complete description of one's instantaneous state of mind should run about the length of one thousand paperback books, which is, I suppose, reasonable. Another way of putting it is to say that a given thought involves arranging a pattern of one thousand book-sized chunks. We might think of a thought as involving a thousand basic concepts, where each concept (like "blue," or "wife," or "car") takes a book to explain:

$$\text{megaplex}^{1000} = (10^{1,000,000})^{1,000} = 10^{1,000,000,000} = \text{gigaplex}.$$

Getting back to the main line of thought, what is the significance of the fact that written language can be coded up by large numbers? One point is that writing is a digital process; writing something down involves creating strings of definite symbols. This is as opposed to speech, which is more of an analog process. The way in which something is said can change its meaning entirely. Skillful writing seeks to digitally mimic these analog effects in various ways.

Another thing that is brought out by the L27 code is the concept of *information.* In terms of pattern or structure, the numerical code for a book has just as much information as does the book itself. The essential information is not so much in the particular representation as it is in the underlying pattern. Usually we just think of numbers as being used for counting things up, but now we see that numbers can also be used to code up information.

Looked at in a certain way, the endless collection of numbers is like a universal library. Each number, if decoded via the L27 method, can be thought of as a certain written text. Conversely, each book can be thought of as an information pattern that corresponds to a certain large number.

Note that there is no limit to how far the coding process can be pushed. Instead of getting the L27 code for one single book, we are also free to stick, say, all 70 million books in the Library of Congress together in catalog order to get an L27 superstring, with a number value some quadrillion digits long.

The Limits of Knowledge

There are infinitely many natural numbers. They surround us in all dimensions like an ocean without shores. Compared to the ocean of number, our whole starry sky, all that is, is less than a germ in the gut of a tube worm warming itself by a volcanic vent at the bottom of the ten-kilometer-deep Mariana Trench.

It is hard for us humans to grasp how small we really are. In biology labs one sometimes sees human brains in formaldehyde. It's strange and frightening to look at a brain, to heft it, and to realize how limited our thinking machinery really is. There are vast realms of phenomena that lie beyond our numbers; there are numbers so big that no one can think of them.

For some reason the *Guinness Book of World Records* does not have an entry for the highest number to which anyone has ever counted, out loud, before witnesses. Given that a day has 86,400 seconds, if someone would take the trouble to name two numbers a second round the clock for six days, they could count up to a million. People have eaten *bicycles* to get in the book, for goodness sakes; can't anyone take the trouble to count to a million?

Of course we don't have to count up to a number in order to think about it. I have a clear idea of what 1,234,567,890 means, even though I haven't counted that high. On a more grandiose level, I also have a clear idea of what the L27 number value of the King James Bible means, although it would be very hard for me to produce the actual base-ten digits of this number. Large numbers code up large amounts of information.

Exactly how much information does a number *K* code up? It takes *K* symbols to write *K* out in unary notation, as a list of *K* strokes, but if we express *K* in base two or base ten, we need fewer strokes. Consider, for instance, a number whose unary description is a list of one thousand two hundred and thirty-seven consecutive ones. To describe this number in the decimal system takes only four symbols:

"1237." To describe the number in the binary system takes eleven symbols: "10011010101."

Recall that, give or take a unit, the log base ten of a number *K* is the number of digits that it takes to write *K* out in decimal form, and the base-two logarithm of *K* is the number of zeros and ones that it takes to write *K* out in binary form.

Since we like to measure information in the simplest possible bits, we will normally say that *the information present in a number K is equal to the base-two logarithm of K.* The base-ten description is not really any more efficient than the base-two description, because understanding a base-ten number requires distinguishing between ten possible digits instead of just two. As was mentioned in "Numbers and Logs," the apparent savings gotten by using base ten is compensated for by the cost of using more symbols. Put more precisely, selecting among ten digits costs 3.32 bits of information, and the base-two log of a number is exactly 3.32 times the base-ten log of a number. That is, the number of symbols in a number's binary notation is 3.32 times the number of symbols in a number's decimal notation. From now on, I will assume that when I write log, I mean log base two. In terms of yes–no bits, I can say that the information in a number *K* is approximately equal to log *K*.

Certain special numbers have very short descriptions, but most numbers have no description much shorter than an actual listing of their digits in one number base or another. "Thirteen to the thousandth power" is a short description of a number that takes some 1100 decimal digits or 3654 binary digits to write out. Most 3654-bit numbers will not have any description shorter than the 3654 bits it takes to actually write them out. These large numbers with no short description may be thought of as small examples of "inconceivable numbers."

I mentioned before that the natural numbers come in four approximate sizes: small, medium, large and inconceivable. The small numbers, once again, can be thought of in the most simple possible way: as collections of dots. Numbers up through about seven are so very small that we can directly visualize them as disordered dot collections. Larger small numbers, like twenty-five, can still be thought of as dot collections, by arranging the dot collection in special ways, perhaps as a square or as a triangle. We understand these dot patterns so well that we can even imagine that the dots are colored black or white in arbitrary ways.

As we move out past a thousand, it becomes practically impossible to visualize numbers as any kind of dot patterns at all. At this point, the coding processes of our numeration system take over. This is the realm of the medium numbers. Instead of thinking in terms of the number of dots that a number like 3721 stands for, we tend to focus on the pattern of its digits. We may be interested to learn that 3721 is 61 squared, but this impresses us as a fact about digits, rather than a fact about dot patterns. The power of the numeration system arises from the fact that a number like 3721 is represented by four pieces of information (the number's four base-ten digits) instead of by a heap of 3721 dots. Given that each decimal digit is worth some 3.3 bits, we see that 3721 involves some 12 actual bits. We might say that a medium number is a number whose base-two log is small. Visualizing the binary name of a medium number is like visualizing a dot collection with some dots marked white (for 1) and some dots marked black (for 0).

If we continue looking for bigger numbers, we soon reach the realm of the large numbers. These are numbers whose binary and decimal representations are too long to be really comprehensible. For a large number, not only is any kind of dot representation out of the question, even the binary representation is barely comprehensible. A number name consisting of a thousand random digits is as good as meaningless. It tells us nothing. It verges on the inconceivable.

Nevertheless, some of the large numbers are significant for us. These are the numbers that have what we might call higher-order names. Thus, for instance, "The L27 number value of *Mind Tools*" is a higher-order name for the number whose L27 code happens to be the string of symbols appearing in the book you are now reading. We might call a name like this a "string name." Names like this are pretty easy to understand if one actually has the book in hand. The book's title becomes a short name for the L27 string that the book's text represents.

A different type of higher-order name is a "mathematical name," a name like "the googolth prime number." Mathematical names are trickier than string names because deciphering them can involve a lot of difficult mathematical work.

Another kind of number name is the "physical name," a name that depends on some aspect of the actual world. A string name works by taking a special kind of physical collection (the letters in a book), ordering it, and treating it as a sequence of digits. A physical name generalizes this process by using things like bodies, or stars, or molecules for the digits. Here are three examples:

The Gender Number. At this moment (13:08:33 EST, January 31, 1986), planet Earth has nearly five billion living human inhabitants. Imagine ordering all these people by age, youngest to oldest. Now define the Gender Number by putting a 1 in the *N*th place if the *N*th person is male, and a 0 if the person is female or of indeterminate sex.

The Galactic Survey Number. Our galaxy has about one hundred billion stars. Imagine that the stars have numbers assigned to them in order of their distance from our sun. Now imagine a hundred-billion-digit-long Galactic Survey Number that codes up information about our galaxy's star systems according to the following code:

0 Exploding star
1 Black hole system
2 Single star with no planets
3 Multiple star with no planets
4 Single star with lifeless planets
5 Multiple star with lifeless planets
6 Single star with planets with life
7 Multiple star with planets with life
8 Star system with nonplanet-based life
9 Artificial star system

As our sun is a single star with planets with life, the first digit of the Galactic Survey Number is 6. If the tenth closest star to our Sun happens to be a double star with lifeless planets, then the tenth digit of the Galactic Survey Number will be 5, and so on.

My Body Number. My body is made up of some octillion molecules. Imagine giving each of my molecules a distinct number name from one through one octillion, and then writing a directory that gives, for each number from one through one octillion, the chemical makeup of the molecule with that number name and the number names of that molecule's nearest neighbors. Now view the directory itself as the L27 code for a number: my Body Number.

Philosophically, these kinds of physical number names seem rather problematic. Take, for instance, the Galactic Survey Number. How, in the teeming whirl of the galaxy, do we actually decide which star is the ten-thousandth closest to us? How large does a rock have to be to be called a planet? What exactly is meant by "life"? And so on. The truth of the matter is that we are not even totally sure if our solar system is really a single-star system or not. Some scientists have recently speculated that our sun may be half of a binary star system, the other star being a very dim and distant object known as Nemesis. Supposedly Nemesis only orbits near Earth every 20 million years or so; the last time it showed up, it caused meteor showers that kicked up enough dust to start the ice age and kill the dinosaurs. This is just a theory, but it's enough to make us unsure whether the Galactic Survey Number starts with a 6 or with a 7! If we can't even figure out the first digit, what right do we have to believe there is such a number at all?

One answer would be that, although we are not in a position to write out the digits of the Galactic Survey Number, nevertheless numbers like it exist as possible information patterns out there in the sea of numbers. Maybe, what with borderline cases, no one number is *the* Galactic Survey Number, but perhaps there is a range of numbers that could serve. This is comparable to the fact that there is no single true text of the Bible; there are, rather, a number of more or less authoritative versions of the Bible, each with its own more or less accurate L27 number.

The same kind of thing holds for the Gender Number. Given that some pairs of people have indistinguishable ages, there will be various more or less correct versions of the Gender Number. No one version of the Gender Number is unassailably correct; nevertheless, some versions of it are more correct than others. That the human race really exists makes us feel that there really are patterns of information that could be identified with the Gender Number.

My Body Number is the least concrete of these physical numbers; the problem here is the basic quantum uncertainty about things like the positions and compositions of molecules. There are probably so many numbers that could qualify as versions of my Body Number that the concept is too vague to be called a number name.

Instead of attacking these definitions for vagueness, one might also attack them for infeasibility. Someone might, for instance, argue that it makes no sense to talk about the Gender Number because no one will ever get around to calculating it. The same argument applies even more strongly to the Galactic Survey Number and to my Body Number. Since it is, for all practical purposes, impossible for any person to ever produce these numbers, we must wonder if it makes sense to say that these are really "names" for anything at all.

It is natural that these names are so *louche*. For I am trying to give examples of names for inconceivable numbers, which are numbers that do not have names at all. Something like the Galactic Survey Number, with its 100 billion random digits, is at the low end of the inconceivable number range.

The high end of the inconceivability scale is the world we live in, viewed as a number. Fine. Now let's rethink it all as space-info.

2

SPACE

Math Space and Real Space

In daily life, the concept of number is tied up with various kinds of power. Putting together a business deal, a machine, or a scientific experiment involves knowing the relative numbers. "Having someone's number" means having control over them, and it is just this kind of number-based control that makes civil libertarians fear the growth of governmental data gathering.

Space, on the other hand, suggests the notion of freedom. We punish criminals by restricting their spatial freedom, by putting them in prison. We free ourselves from ignorance by expanding our minds with books. Our language is so digital and concept-oriented that it is hard to talk about pure, empty space. Space has no inherent locations. This means that instead of suggesting static patterns, space suggests *possibilities of motion.* Our real knowledge of space comes not from reading, but rather from moving around. An athlete knows space in the same intimate way that an accountant knows numbers.

Speaking quite generally, any kind of motion at all suggests an underlying space. We regard our bodies as moving in three-dimensional physical space, but we also think of the nation's economy as moving in multidimensional space, a space with axes for GNP, inflation, unemployment, etc. One's emotional state can be regarded as moving about in some kind of mood space. Generally, any sort of change suggests the idea of space.

Abstracting the notion of counting various types of collections leads to the standard mathematical numbers. By the same token, looking at the bare idea of smooth change leads to the concept of mathematical space.

Mathematics discusses a great variety of lovely spatial forms. There are numerous ways of classifying these forms. One method is to organize space forms according to their intrinsic dimensionality. A point is called zero-dimensional. A *curve* is called one-dimensional, even if the curve is bent in some higher-dimensional space. A surface, curved or not, is two-dimensional. There are various varieties of space that we call three-dimensional. Four-dimensional forms are called hyperspaces, and certain infinite-dimensional forms are called Hilbert spaces.

The simplest space forms of dimensionality one through three are as follows: lines and circles, planes and spheres, flat spaces and hyperspheres. (A hypersphere is a certain curved 3-D space.)

Mathematics has a great variety of names for specific space forms of dimensionality one. That is to say, mathematics has studied lots of different curves. Some well-known curves, in roughly increasing order of complexity, are lines and polygons; conic sections, cubic curves, quartic curves, and algebraic curves of higher degree; sine curves and exponential curves; spirals; helices and curves in higher space; branching curves. I will describe most of these curves in this chapter, but before starting, I want to say a bit about the ways in which mathematical space differs from real space.

But what *is* real space? We commonly think of the world as three-dimensional; by this we mean that our gross body motions have three degrees of freedom — left/right, forward/backward, up/down — but even this "real" space is a kind of abstraction. What I actually *know* about the world is my immediate perceptions, which cascade in, pell-mell, as my head and body move this way and that. I know that certain kinds of body movement (such as walking forward) will change my visual space perceptions in certain predictable ways (such as changing a building corner's apparent angle). My mind unconsciously integrates a large amount of such information to produce an internalized 3-D model of the objects in my vicinity. This internal model-building is by no means a trivial process — one of the great problems in contemporary robotics is to write programs that can convert coupled 2-D pictures and records of motion into 3-D models.

We do not live in true mathematical space, but mathematical space serves as a convenient idealization of physical space. Plenty of things in our world look *pretty much* like straight lines, or *pretty much* like parabolas, or *pretty much* like spheres. The usefulness of geometry arises from the fact that we can take a real-world shape, approximate it by an idealized mathematical form, draw certain logical conclusions about the mathematical form, and then, finally, translate our conclusions into some predictions about the original real-world shape. In our daily experience, Earth is nothing at all like a smooth sphere, yet if we know the length of the equator and wish to estimate the Earth's volume, it is useful to approximate Earth by a sphere and use a mathematical formula that gives a sphere's volume in terms of its circumference.

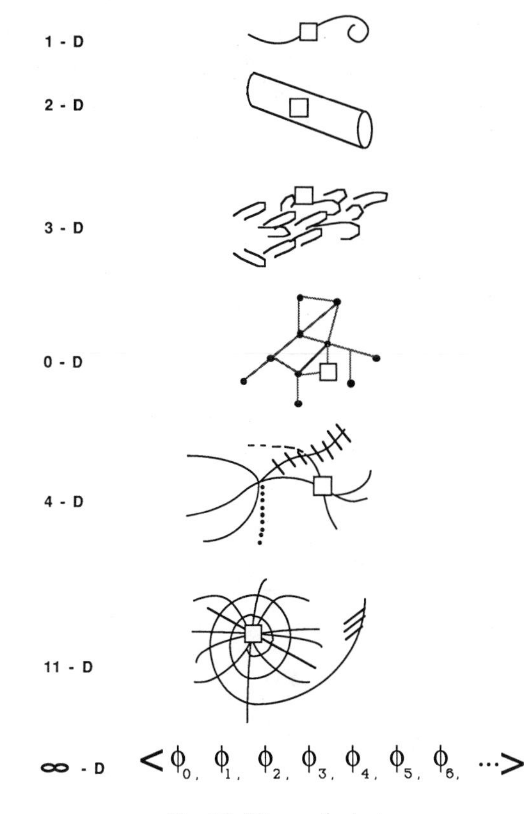

Fig. 33 Views of a hair.

What is it that keeps our physical world from corresponding *exactly* to the space forms of mathematics? I think the key problem is the notion of "dimensionality." Dimensionality is not a clear-cut notion in the physical world. To the naked eye, a human hair looks like a dark line, a one-dimensional curve. Under a magnifying glass, the same hair looks like a smooth tube, a two-dimensional surface. Under a microscope, the hair becomes a three-dimensional pile of scaly scallops. Classical physics says the hair is "really" an arrangement of pointlike particles in 3-D space. Relativistic physics says the hair is "really" a bumpily curved region of hyperspace. The recently revived Kaluza–Klein theory says matter is eleven-dimensional. Quantum physics says that the hair is a wave-function pattern in an infinite-dimensional Hilbert space. The problem of physical dimensionality is dual: there is no constancy at the perceptual level nor is there any universally agreed-upon constancy at the theoretical level.

Compare the hair to a truly one-dimensional curve — a straight line. A line — or a plane, or a flat space — is characterized by its

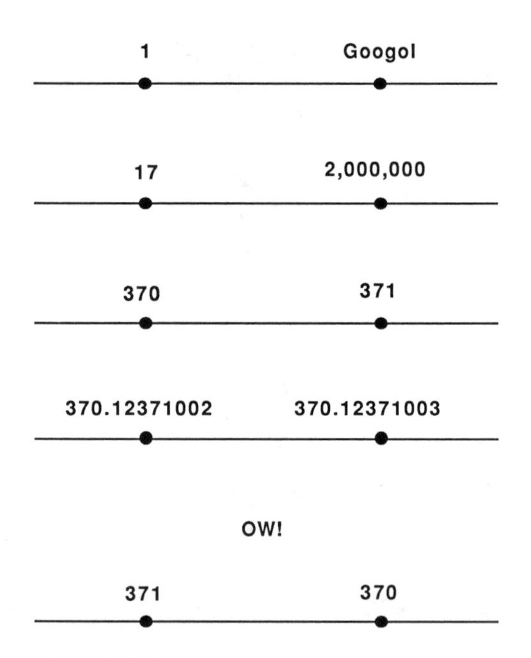

Fig. 34 A line looks the same at every size scale.

homogeneity, its lack of distinguishing features. A mathematical line looks the same at every magnification. If you were falling toward a line, and if you were somehow able to see it, you would have no way of telling when you were about to hit. The line would look the same right on up to when it cut your eye in half.

Our troubles with specifying the dimensionality of real-world objects come in part from the world's inherent complexity. Later in this chapter we will see that mathematics has quite recently developed a variety of forms that have the same complexity and vague dimensionality as real objects. These odd forms are called *fractals.*

First, though, we will look at some of the standard mathematical curves suggested by the world around us. It has been said that there are six main kinds of simple curves to be found in nature: polygons, circles, waves, spirals, helices, and branchings (see Fig. 35). We will look at examples in the sections to come.

Tiles, Cells, Pixels, and Grids

In the most general sense, a polygon is a figure consisting of a finite number of straight line segments joined end-to-end in such a way that the tip of the last segment touches the tail of the first segment. A general polygon may cross itself, or it may look caved in.

Usually, when we talk about polygons, we prefer to talk about the *regular* polygons. A regular polygon has all its sides the same length and all its angles the same size. It bulges out all around, and it doesn't cross over itself. There are infinitely many regular polygons: triangle, square, pentagon, hexagon, heptagon, octagon, nonagon, decagon, etc. Given any value of N, we can hinge together N equal line segments and circle them around into a regular N-gon.

When we are talking about regular polygons, we are still as much in the domain of number as in the domain of space. Thinking about, say, a decagon is not so very different from thinking about the number ten. A simple computer graphics program to draw an N-gon includes little more information than the number N itself. In the computer

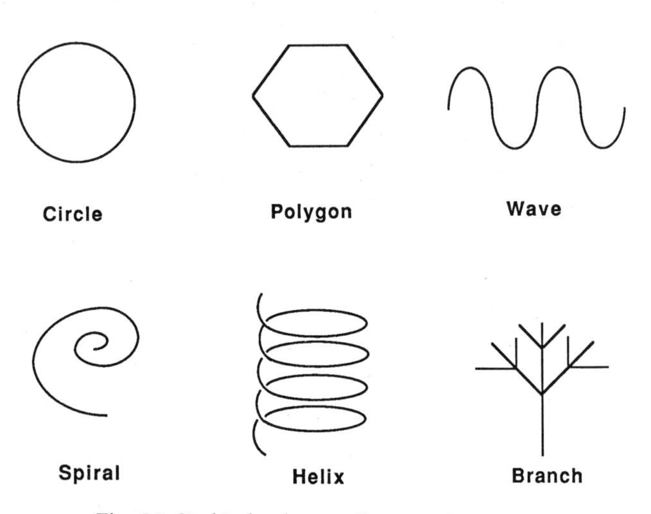

Fig. 35 Six kinds of naturally occurring curves.

language LOGO, such a program would take the form "Repeat *N* Times: [Go forward one inch, turn left 360/*N* degrees]."

The regular polygons begin to seem a bit more space-like if we start to look for ways in which the plane can be covered with them with nothing left over. Such a covering of the plane is called a *tessellation,* which is a very nice-sounding word. It comes from the Latin *tessera,* meaning a small cube, a die, or a square-faced tile.

The only three fully uniform tessellations use only one kind of regular polygon each; we can tile the plane with triangles, with squares, or with hexagons.

The hexagonal tessellation is often seen on bathroom floors, probably because the absence of any continuous straight lines through it makes cracking less likely. If one looks at a mass of foam, such as soap suds in the sink, one will notice that the bubbles fit together into a pattern roughly resembling the hexagonal tessellation. This is a result of the fact that the bubbles' soap films are under tension and try to be as small as possible. The hexagonal tessellation is the one repeating pattern that uses the least number of lines to cover a given area, so as each soap film tries to get as small as possible, a mass of bubbles naturally takes on the shape of a bunch of hexagons.

For a slightly more complex group of tessellations, suppose we

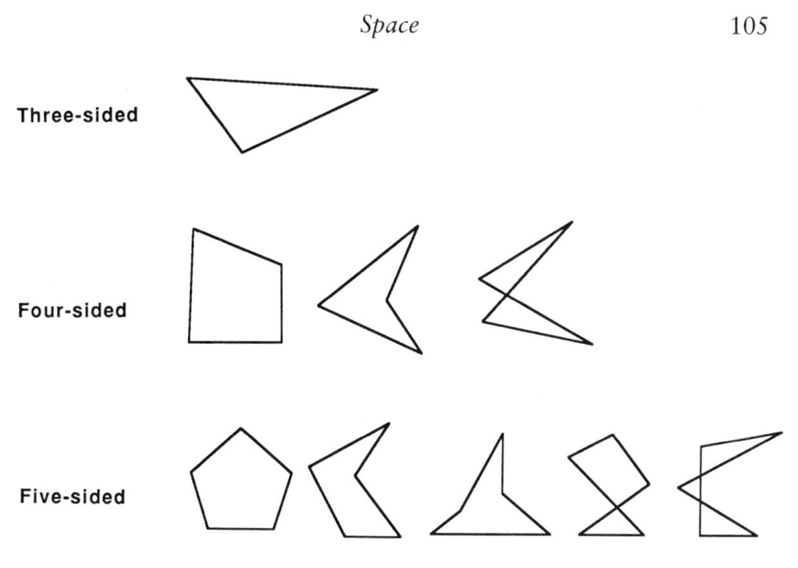

Fig. 36 General polygons.

allow more than one kind of polygon, but require that each corner look like every other corner — require, that is, that each corner be the meeting point of the same number of polygons of the same shape as those meeting at any other corner. Such tessellations are called *homogeneous,* and are found only in the finer bathrooms.

If we give up the requirements that each tile be a regular polygon and that each corner look the same, a great many strange tessellations

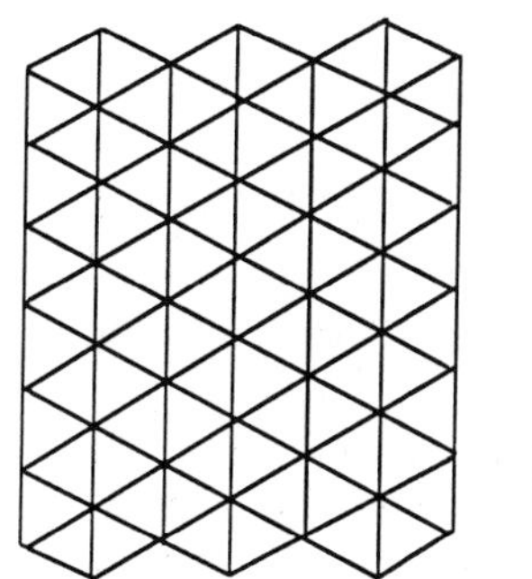

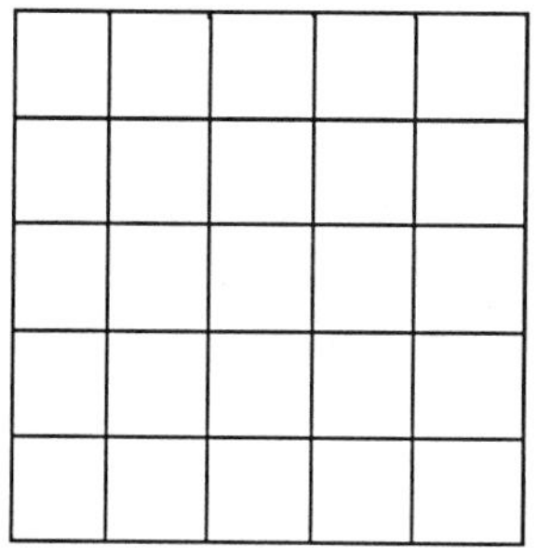

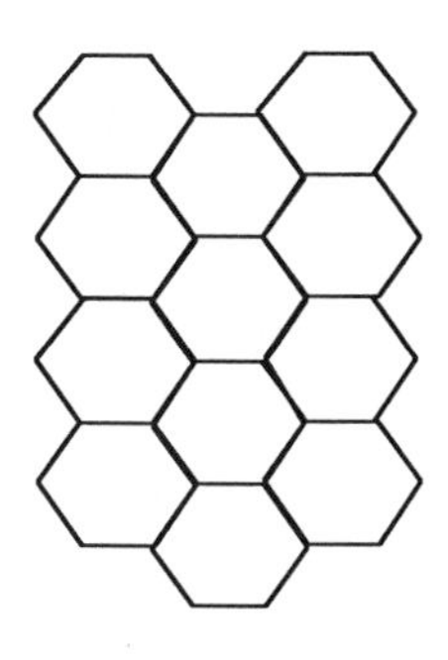

Fig. 37 The three simplest tessellations.

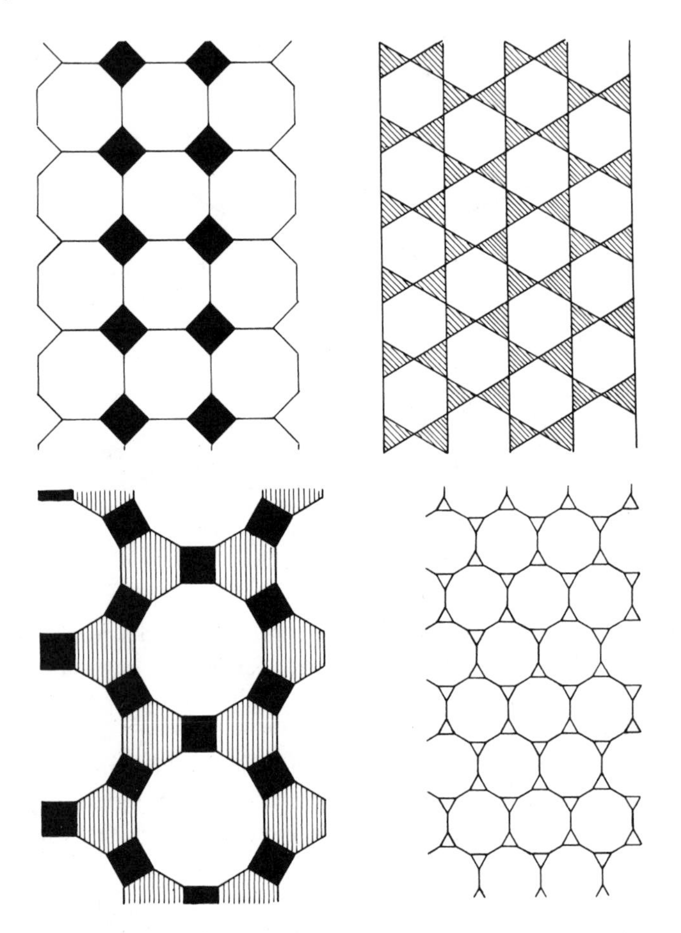

Fig. 38 Four homogeneous tessellations. Drawings from Hugo Steinhaus, *Mathematical Snapshots,* Oxford University Press, 1950.

can be found. One very attractive one is made of irregular pentagons and is often used as a cobblestone pattern in Europe and the Near East.

A tessellation is said to be *periodic* if there is a basic block of units in it that repeats over and over like wallpaper. More precisely, a periodic tessellation is one on which you can outline a region that repeats over and over without rotating. All the tessellations we have looked at so far are periodic. In each there is a smallest region (possibly made of more than one tile) that generates the whole pattern by periodic repetition.

Once you have seen a big enough piece of a periodic tessellation, there are no surprises left. A simple example of a nonperiodic tessellation can be gotten by arranging triangles in concentric 12-gons.

This pattern is predictable in the sense that it is all rings of triangles,

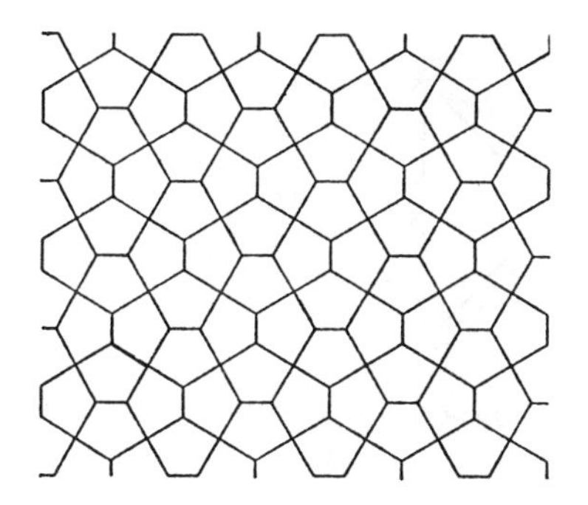

Fig. 39 Tessellation with irregular pentagons.

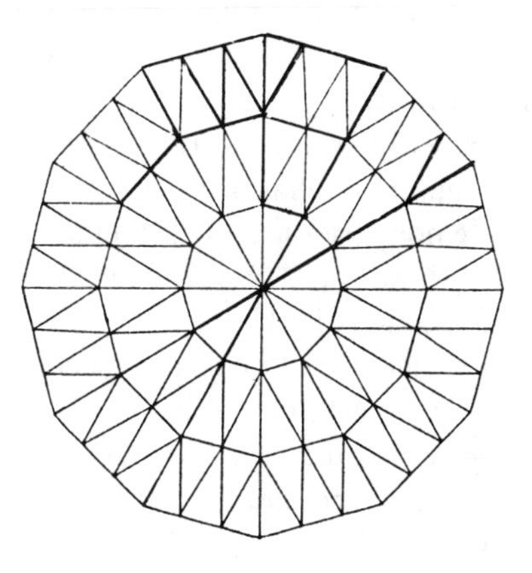

Fig. 40 A predictable, yet nonperiodic tessellation. Drawing from Martin Gardner, "Mathematical Games," *Scientific American*, January 1977.

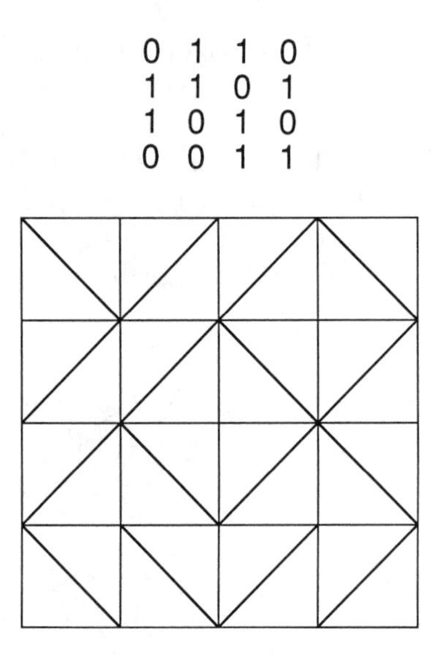

Fig. 41 An irregular grid of zeros and ones codes up a nonperiodic tessellation.

but we say it is nonperiodic because there isn't any way to get the whole thing by just repeating one piece of it left and right and up and down, without rotation. A less predictable nonperiodic tessellation can be gotten if we imagine filling an endless two-dimensional grid with random zeros and ones. Now, in place of each zero, put a square diagonally cut in two by a down-sloping line, and in place of each one, put a square cut into two triangles by an up-sloping line. This leads to an irregular tessellation based on right triangles.

A very remarkable nonperiodic tessellation of the plane was found in 1974 by the British scientist Roger Penrose. Penrose, who is a professor of mathematics at the University of Oxford, does most of his work in relativity theory and quantum mechanics. He is perhaps best known for his early work on the theory of black holes; in recent years he has been working on a new theory of matter called "twistor theory." As if that weren't enough, he spends his spare time thinking up strange, mind-warping mathematical games.

Penrose's tessellation is based on two special tiles. Note that each tile is drawn with two black dots. To prevent the tiles from lining up

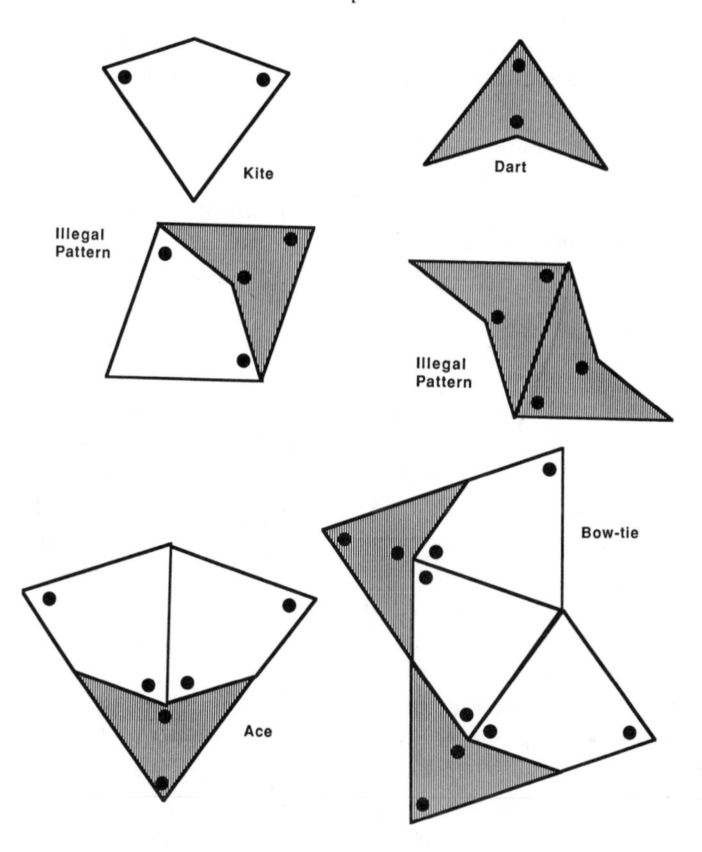

Fig. 42 Penrose tiles.

in any orderly pattern, Penrose makes the rule that the tiles must always be placed in such a way that each corner with a dot touches only corners with dots. The big tiles are called "kites," and the little ones "darts." Various small Penrose tile patterns have names like "ace," "short bow tie," and "batman."

The nomenclature is due to John Horton Conway, a bizarre and brilliant professor of mathematics at the University of Cambridge. Weird, colorful terminology is a Conway trademark, as is the buzzing, headachy feeling one gets when trying to understand him. Working together, Penrose and Conway proved that there are infinitely many different ways to cover the whole plane with the kite and dart tiles.

What makes the Penrose tessellations so remarkable is that *none of them is periodic.* As you continue out through any Penrose tessellation, you encounter, more or less randomly, every possible finite pattern into which the tiles can be formed. The simple tile shapes (based on angles of 36 and 72 degrees) somehow force the whole infinite plane into a nonperiodic pattern. This is a little like seeing a very simple computer program print out an endlessly complex message.

Recently I made myself 320 Penrose tiles — 200 kites and 120 darts. I did this by tracing a lot of kites onto one sheet of paper and a lot of darts onto another sheet of paper. I photocopied a bunch of copies of these master sheets, and then got myself some rubber cement and three large sheets of posterboard, one red and two yellow. I glued the copied kite blueprints to the back of the yellow posterboards and glued the dart blueprints to the back of the red. I got scissors and cut out yellow kites and red darts. I put black dots on them to keep from putting them together illegally, and then I tried to get them all to fit together. After two or three days of frenzied tile-shifting, I got something like Fig. 43.

Fitting the Penrose tiles together is surprisingly hard. Often one works out to a point where no more tiles will fit legally; then one has to backtrack, remove the last dozen or so tiles, and try again. For all I know, the design I show here cannot be continued outward without some changes. Not too much is known about the Penrose tiles, and it may turn out to be theoretically impossible to find a general technique for predicting which starting patterns can indeed be extended to a full tessellation of the plane. This "Penrose tile problem" is probably equivalent to the unsolvable "Turing halting problem," which will be discussed in chapter 3.

What makes the Penrose tiles important enough to make it worthwhile torturing yourself with them is that they serve as a kind of model for how our universe might be built up. It is conceivable that our universe is, at some deep level, made up of many, many copies of a few very simple components that can only be linked together in certain definite ways. The surprising lesson of the Penrose tiles is that one can have simple rules and components that necessarily combine to make a vastly complicated whole!

This basic insight underlies another of John Horton Conway's creations, the game of Life. Conway's Life is not really a "game" in the sense of "contest"; it is, rather, a sort of solitaire, an activity a bit like watching a kaleidoscope. One starts with a plane tessellated into squares, and one shades in some of the squares. The squares are called "cells," the shaded cells are called "living," and the clear cells are called "dead." If we count touching at one corner, then each cell is in touch with eight neighbors. Now the following transition rule is simultaneously applied to each cell: If a vacant cell has exactly three living neighbors, then these neighbors "reproduce" and the vacant cell becomes occupied. If a living cell has two or three living neighbors, then it is happy and stays alive. If a living cell has less than two living neighbors, then it dies of loneliness, and if a living cell has more than three living neighbors, then it dies of overcrowding.

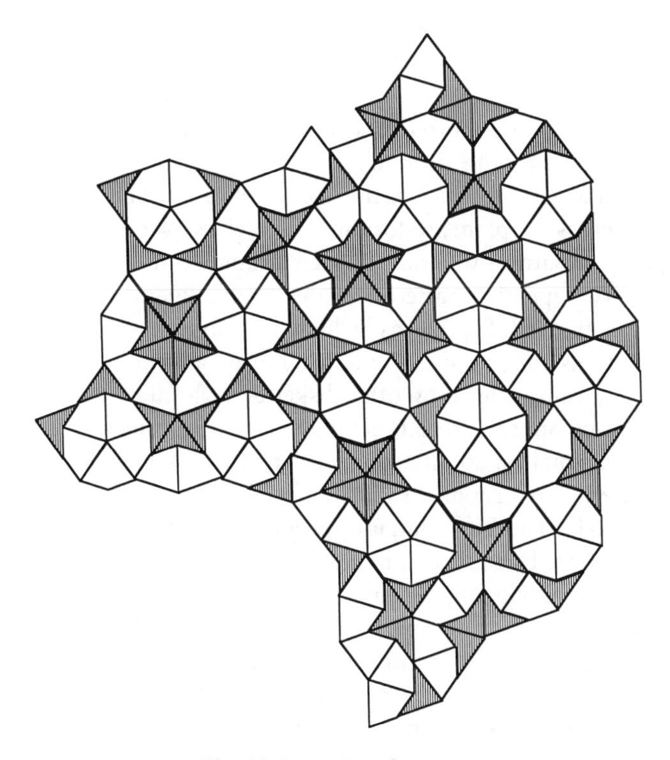

Fig. 43 Part of a Penrose tessellation.

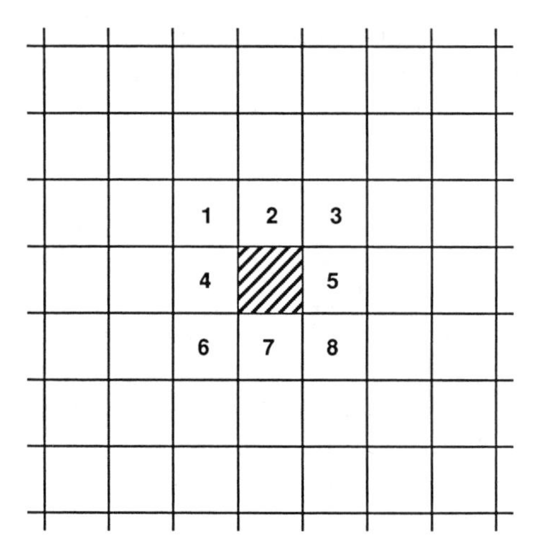

Fig. 44 A marked cell and its eight neighbors.

If you start with a more or less random pattern of living cells and apply the transition rule over and over, you see many curious patterns emerge. Some patterns, such as the one known as the "glider," manage to move across the plane, re-forming themselves over and over. More commonly, a pattern will run through a number of stages and reach a stationary, repeating pattern like the "traffic light."

One can evolve Life patterns by hand on graph paper or on a "Go" board, but it is immensely easier to run Life on a computer. A number of software companies now offer diskettes with Life programs. Any image on a computer screen is actually made up of tiny square "tiles" of light. The smallest dot that a screen can show is known as a "pixel." Most screens have room for something like 200 × 300, or 60,000 distinct pixels. The Life computer programs treat the pixels as individual cells of a Life game; a living cell is lit, and a dead cell is dark. By using a randomizing function, one can cover the screen with a "primordial soup" of on and off bits, and then let the rules of Life take over. If the transformations are done rapidly, say sixty times a second, the screen fills with frenzied activity. Dots race around and clump into things like worms. The worms crawl around the screen, colliding and eating each other, casting off bits of stable debris.

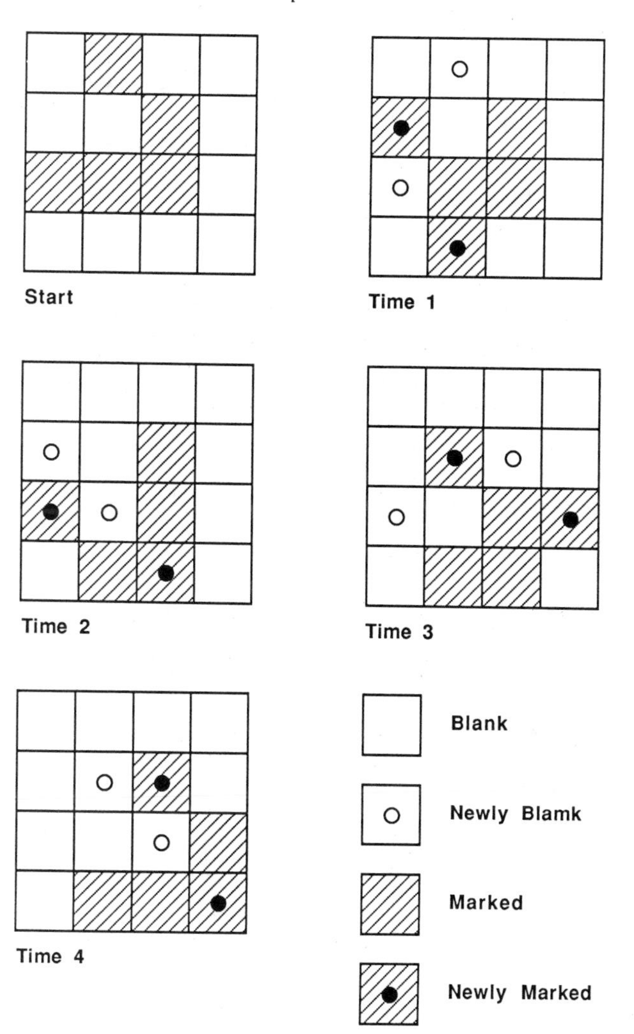

Fig. 45 The glider pattern in action.

Some patterns, which Conway calls "glider guns," pump out endless streams of gliders.

The game of Life is an example of a general class of processes known as "cellular automata." The game of Life is said to be cellular because the same transformation rule is applied at each of the cell

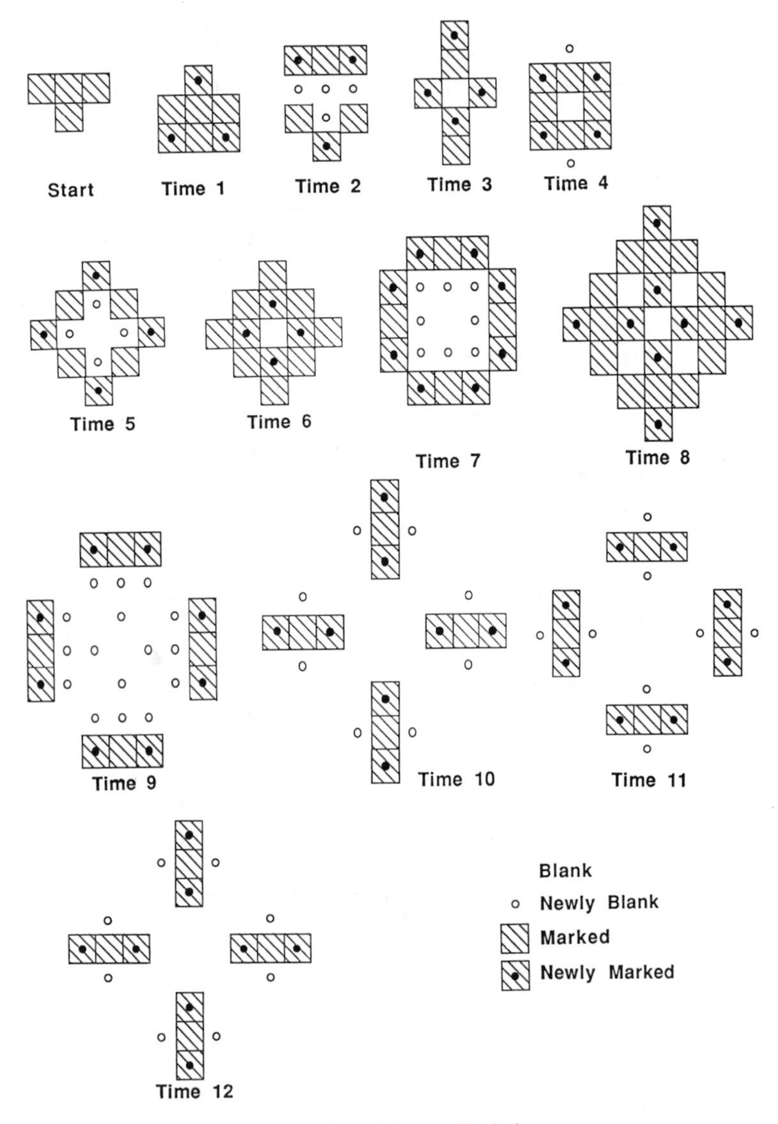

Fig. 46 From T to traffic light.

sites; it is like an automaton, or computer, in that it accepts data input in the form of the starting pattern of marked cells, it changes this data in accord with definite rules, and it gives you data output in the form of the new pattern of marked cells. A generalized cellular automaton can be thought of as a large array of independent, identically programmed computers, one computer per cell. Each cell can be colored any one of some specified colors. Life uses just two colors, black and white, but any finite number of colors is allowable. With each tick of the clock, each cell looks at its neighbors and decides what color to display next. The transition rule can be more complicated than the rule for Life, but if there are enough cells, very complex results can be gotten with rules that look quite simple.

The notion of a cellular automaton was first investigated by Stanislaw Ulam and John Von Neumann in the early 1950s. Von Neumann was interested in the notion of a machine that can reproduce itself. Before Von Neumann's work, such a notion had seemed ludicrous, even blasphemous. Our industrial machines produce things much simpler than themselves, and many people had thought that some mysterious extra-scientific "vital force" is involved in true self-reproduction. Von Neumann was able to show, however, that it *is* possible to have

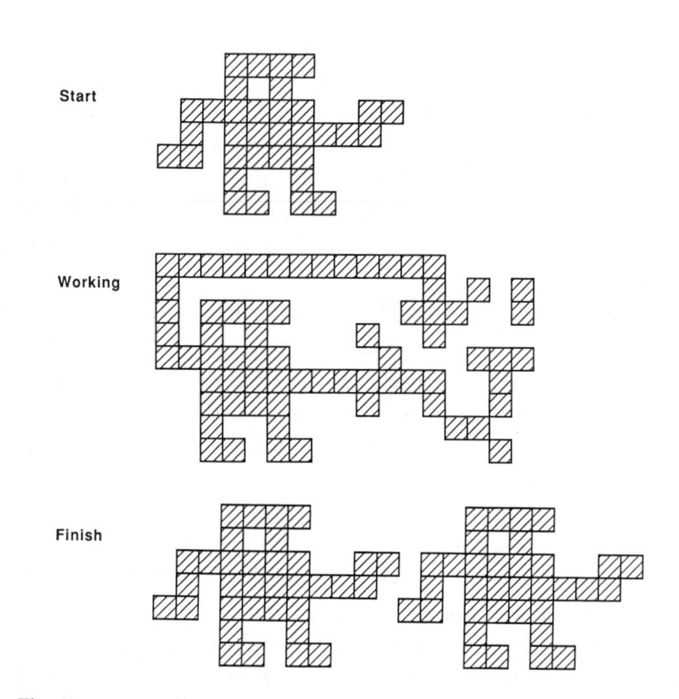

Fig. 47 How a cellular automaton pattern might reproduce (greatly simplified).

robots that build factories to produce new robots just like themselves. Ulam, coinventor of the hydrogen bomb, showed Von Neumann how to simplify his argument by thinking of his robots as cellular automata — as patterns, that is, on a large grid of square cells. A plane filled with square cells becomes a simplified version of physical space. A pattern in such a plane, evolving according to some fixed transformation rules, is like a machine or a robot. The question of mechanical self-reproduction then becomes the question of whether any cellular automaton patterns copy themselves. Shortly before his death, Von Neumann came up with a description of a cellular automaton world in which certain patterns will indeed build copies of themselves. Later, Conway showed that this is also true in the game of Life. There is a (very large) starting pattern of lit cells which will slowly send more and more copies of itself out across the Life plane.

Aside from Conway, one of the next early investigators of cellular automata was Edward Fredkin, who in the 1960s formulated a concept of "information mechanics," in analogy to quantum mechanics. Starting from the basic premise that the world is made of information, Fredkin proposed modeling the entire physical world as a very large three-dimensional cellular automaton. This is a modern version of an idea of Gottfried Leibniz's: The world is made up of indivisible "monads," each of which acts independently. The gross, large-scale phenomena we observe are statistical averages of the harmonious workings of zillions of monads. According to Fredkin's vision, our world may consist of tiny three-dimensional monads that update their states some decillion times a second.

One obvious difficulty with such a computerized world view is that the world does not in fact look as if it is made up of pixels. One feels that, although some of the world's information is digital, much of it is also analog. Things like heat, loudness, and distance seem to increase smoothly, rather than in digital jumps.

It could be, though, that we simply aren't able to examine the world closely enough to see its digital texture. Something feels analog to us if its pixels are too small for us to readily sense them. Put a bit differently, a process seems analog if its digital representation involves an inconceivably large number of separate bits. In a mosaic mural, a pixel is the size of the tiles being used. A pixel on a computer screen is a dot of minimal size that can be either light or dark. Perceptually, a pixel is a spot so small that you can't be sure if you see it or not.

When we are acting most naturally — running, looking around, talking to our friends — we are carrying out processes that are near the limits of our brain's information processing capacities. So many digital bits are involved that there is no thinking space left in which to "step aside" and see the bits. Our natural activities therefore feel like analog processes that cannot be modeled digitally.

Nevertheless — and this point is worth stressing — any physical process can be modeled up to any desired degree of precision by a complicated-enough digital computation. The difference between a digital choice like which word you say and an analog choice like which tone you speak in is a matter of degree. There are so many possible tones to speak in that we despair of viewing them as settings on some digital switchboard. This is simply a failure of the imagination, however, not a reflection of an underlying reality. For, consider that there are, after all, only some million different ways in which to say a word, and a million is about two to the twentieth, so one's tone of speech, which initially seems like a totally analog concept, can be expressed by an additional twenty digital bits per word.

Most of us must be familiar with the way in which mathematicians correlate decimal numbers with points on a line. Given a line, one randomly chooses a point on the line and calls this point the "origin." The origin is often labeled 0. Once the origin is chosen, a size scale is determined by randomly choosing a second point, labeled 1. The line segment between 0 and 1 is called "the unit interval." By laying out copies of the distance between 0 and 1, both to the left and to the right, one gets points labeled 2, −1, 3, −2, 4, and so on. Dividing each interval into ten equal pieces gives us points with names like 2.1, 2.2, 2.3; redividing each subinterval into ten smaller pieces gives us points with names like 2.31, 2.32, 2.33; and the process can be continued as long as one is able. These number names are usually just called "decimals," but it is more correct to call them "decimal point fractions."

In practice, a name like 2, or 2.3, or 2.37 is not really the name of an exact point location on the line. If all I know about a point's location is the number 2, then the point can be anywhere in the unit interval between 2 and 3. By the same token, to say a point's location is 2.3 can be thought of as saying the point lies somewhere in the interval between 2.3 and 2.4. If I wish to express greater precision, I can give the name 2.30, which now refers to an interval one-hundredth unit long. Any degree of measuring precision sets up a kind of grid — a level of accuracy. If by some miracle we would reach infinite accuracy, then we would end up with endlessly long decimal names (like 2.37583200917464 . . .) for infinitely small intervals. In this idealized situation we would have a kind of number name for every point on the line; this would be an example of infinity providing a bridge between number and space. Mathematicians often pretend that just such an idealized situation is the case, but the actual fact is that we never measure anything to more than finitely many places.

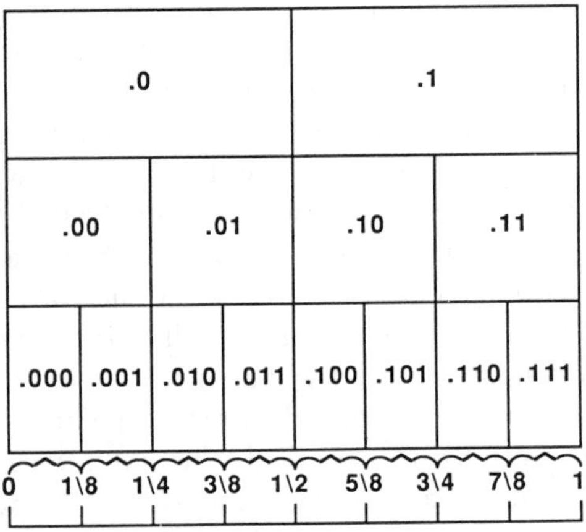

Fig. 48 Binary point fractions.

To make this clearer, we can use "binary point fractions" to express numbers between 0 and 1, with the understanding that a 1 in the *N*th place to the right of the point stands for one over two to the *N*th power. That is, $.1_{\text{TWO}} = 1/2$, $.01_{\text{TWO}} = 1/4$, $.001_{\text{TWO}} = 1/8$, $.1011_{\text{TWO}} = 11/16$, and so on. A binary point fraction that starts ".0" lies in the left half of the unit interval, a binary point fraction that starts ".01" lies in the right quarter of the left half of the unit interval, and so on. Each additional bit doubles the precision.

When one is actually sketching things, one is in fact more likely to use binary point fractions than the more familiar decimal point fractions. The reason for this is that it is easier to divide an interval in half than it is to divide it into tenths. (See Fig. 49.)

Using a pencil to subdivide a line as much as possible, I reach a limit based on the fact that I really can't distinguish pencil marks that differ by much less than a millimeter. In practice, then, the "analog" process of picking a point on a ten-centimeter line is really a choice among 100 possibilities, which means something like six or seven digital bits. Put differently, an "analog bit" — viewed as a location on an interval — is in practice worth about seven digital bits. (See Fig. 50.)

Still, to try to see everything digitally is sometimes to miss the point. This becomes very clear if we compare the smooth arc of a parabola in space to the clumsy digital staircase of pixels that represent it on a computer screen. In the next section we'll forget about digitization and enjoy some curves for what they really are.

Algebraic Curves

The standard unit of digital information is a digital bit, a choice between 0 and 1. In talking about space curves it is convenient to think of the standard unit of analog information as being a location on a line. When I talk about "a piece of analog information," I am going to mean a location on a line. In the last section we saw that, in practice, such an analog bit is worth about seven digital bits, although if one were concerned with close tolerances and very high degrees of precision, an analog location would code up a correspondingly larger number of digital bits of accuracy. But let's set digital bits aside for now, and think of a point on a line as being one piece of analog information.

In the seventeenth century, the French mathematicians Pierre Fermat and René Descartes hit on the idea of extending the number naming system from the line to the plane. This idea developed into the now-familiar notion of using "cartesian coordinates" to specify

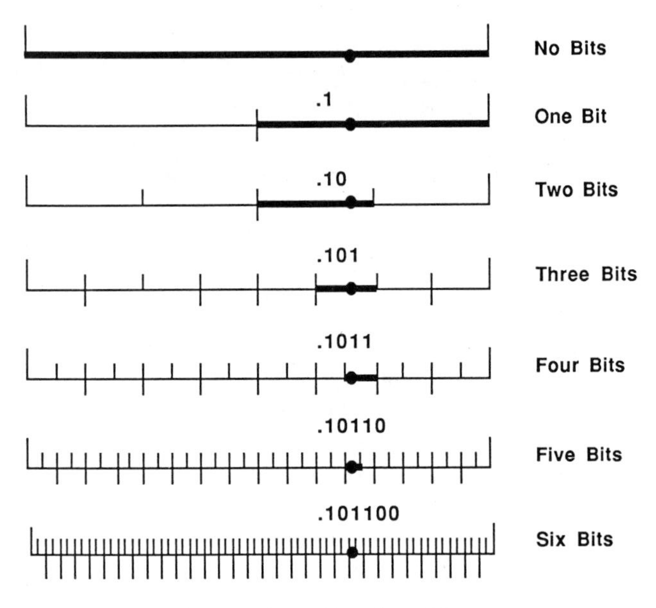

Fig. 49 Binary bits of precision.

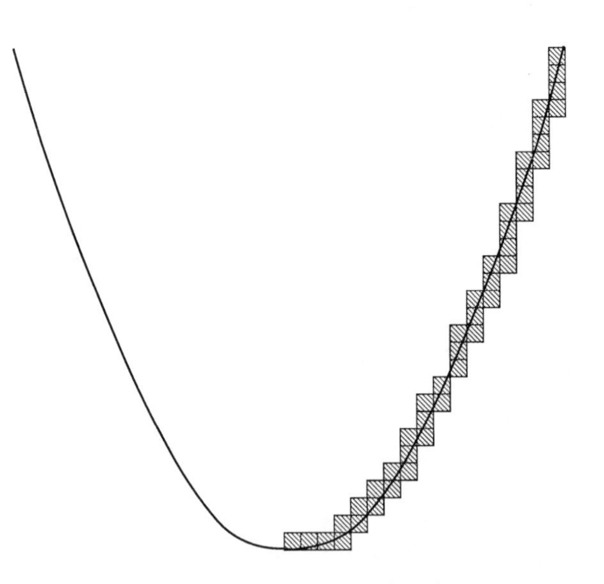

Fig. 50 Analog versus digital.

locations in the plane. Each point on a curve can be thought of as making a match between its horizontal distance from the origin (its x coordinate) and its vertical distance from the origin (its y coordinate). If a point on a line codes one analog bit x, then a point in the plane codes two analog bits (x,y).

The great value of this approach resides in the fact that a given plane curve can often be described in terms of an algebraic relationship between x coordinates and y coordinates. Here we see the logic of algebra providing a bridge between space curves and pairings of brute numbers.

Not every plane curve corresponds to a simple algebraic equation, but many of them do. These are the algebraic curves. Formally speaking, an algebraic curve is a curve that is given by an equation involving whole-number powers of x and y. For the rest of this section, we will look only at algebraic curves. This is not really so severe a limitation as one might think. Any smooth squiggle at all can be approximated by an algebraic curve.

Algebraic curves are grouped according to the complexity, or degree, of the equations describing them. The degree of an equation is

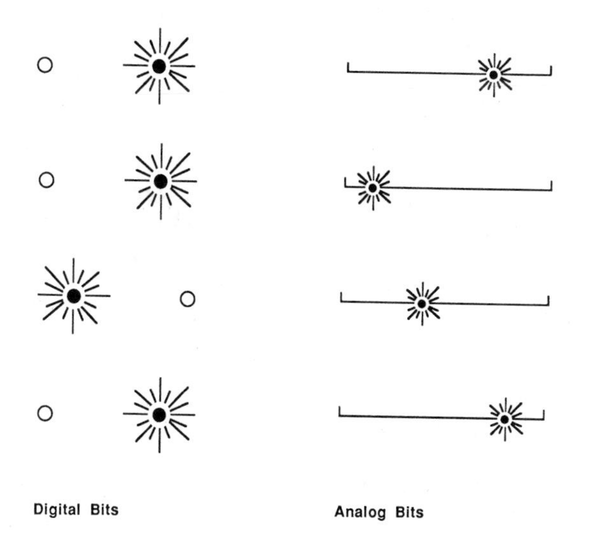

Fig. 51 Two kinds of information.

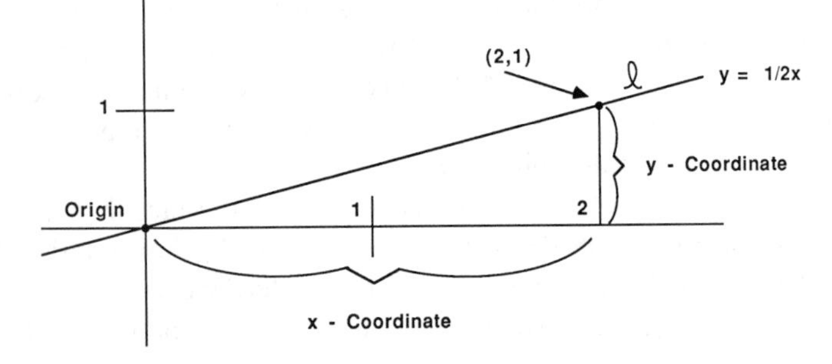

As Number - The ratio 1 to 2.
As Space - The curve ℓ.
As Logic - The equation y = 1/2x .
As Infinity - The set of all number-pairs giving coordinates of points on ℓ.
As Information - Two analog bits, worth about seven digital bits each.

Fig. 52 Notions of a line.

the highest combined power of x's and y's that occurs in any one term:

$8x - 5y = 2$ is an equation of the first degree;
$x^2 + y^2 = 100$ is an equation of the second degree;
$x^2y - y^2 + xy - x = 1$ is an equation of the third degree;
$x^4 - x^2y^2 + 200xy^3 = 0$ is an equation of the fourth degree.

All first-degree equations describe straight lines, and all straight lines have first-degree equations. Most of us learn all about this in college algebra, or in high-school analytic geometry, where students practice converting first-degree equations into lines and lines back into first-degree equations. $Ax + By = C$ is a general equation of the first degree.

It is interesting to ask how many "parameters," or independent pieces of analog information, a first-degree equation involves. At first it looks as if the general first-degree equation involves three parameters: the x coefficient A, the y coefficient B, and the constant term C, but C can always be taken to be 0 or 1. Either C is 0 (because the line goes through the origin), or it is not 0. If C is not zero, then we divide each term of the equation by C to get 1 on the right side.

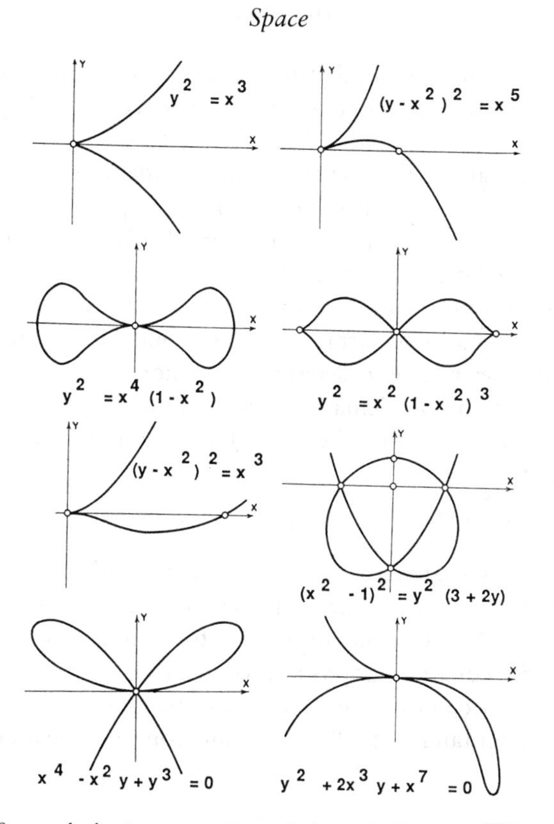

Fig. 53 Some algebraic curves. From Robert C. Yates, *A Handbook on Curves and Their Properties*, J. W. Edwards, Ann Arbor, 1947.

$$\begin{aligned} Ax + By &= C & 8x - 5y &= 2 \\ (A/C)x + (B/C)y &= 1 & 4x - 2.5y &= 1 \\ A'x + B'y &= 1 & (4)x + (-2.5)y &= 1. \end{aligned}$$

There are really only two independent coefficients in a first-degree equation, so this family of equations is sometimes called a "two-parameter family." Given the correspondence between first-degree equations and straight lines in the plane, we see that the lines are also a two-parameter family. This means that two pieces of analog information specify a line. The two parameters might be the x intercept and y intercept, or the slope and y intercept. That two points determine a line is related to the fact that lines are a two-parameter family. In general, N points determine a unique member of an N-parameter family.

Now, what about the algebraic curves with second-degree equations? It turns out that these equations correspond to the familiar shapes circle, ellipse, parabola, and hyperbola. These algebraic curves are collectively known as conics, because they can all be produced as the intersection of a cone and a plane. We just saw that two pieces of analog information determine a straight line. How many pieces of analog info does it take to determine a conic?

Keeping in mind our remark above about dividing through by the constant term, we see that the general xy equation of second degree has the form

$$Ax^2 + Bxy + Cy^2 + Dx + Ey = 1.$$

As with the first-degree case, it is also permissible to have 0 on the right, in the case that the curve in question goes through the origin.

Every conic has an equation of this form and, conversely, every second-degree equation graphs as a conic. There are five coefficients in the general second-degree equation, so this family of equations is a five-parameter family. It follows that any conic section is determined

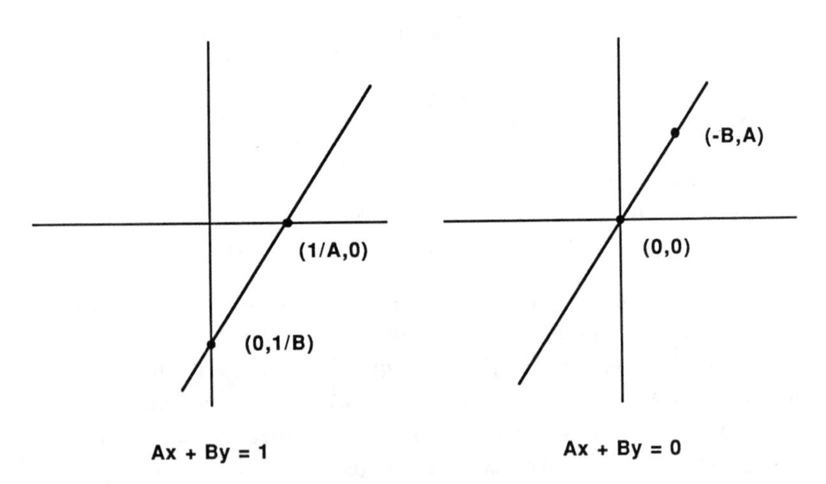

Fig. 54 Graphs of general lines.

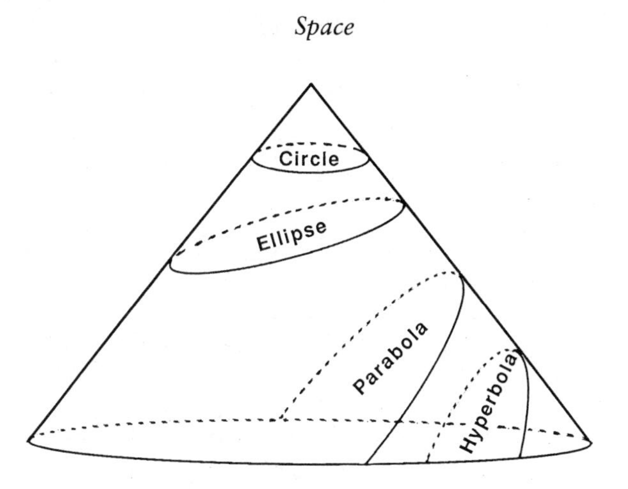

Fig. 55 Conic sections.

by at most five analog bits of information. To draw a general ellipse, for instance, we need to specify (1) the x coordinate of the ellipse's center, (2) the y coordinate of the ellipse's center, (3) the angle that the ellipse's long axis makes with the horizontal, (4) half the length of the ellipse's long axis, and (5) half the length of the ellipse's short axis. Just as any two points in the plane determine a line, any five points in the plane determine a conic. As long as no four of the points lie on a straight line, the conic is *uniquely* determined.

It is interesting to think about how many ways the overall shapes of various kinds of curves can vary. Consider the ellipse. If we don't care about where the ellipse is located, or how it is oriented, we see that there are really just two ways in which an ellipse's shape can vary: Its overall size can change, and its *oblateness,* or flatness, can change. If we think of scale and oblateness as being independent axes in an abstract "parameter space," we get a two-dimensional space, each point of which corresponds to a particular ellipse.

A parameter space is a very good way to organize a family of curves. Note that the ellipse's two-dimensional parameter space includes as a subspace the one-dimensional parameter space of all possible circles. As long as one doesn't care about position, only one thing about a circle can change: its scale, or radius.

It is not so well known that the parabolas also fit into a one-

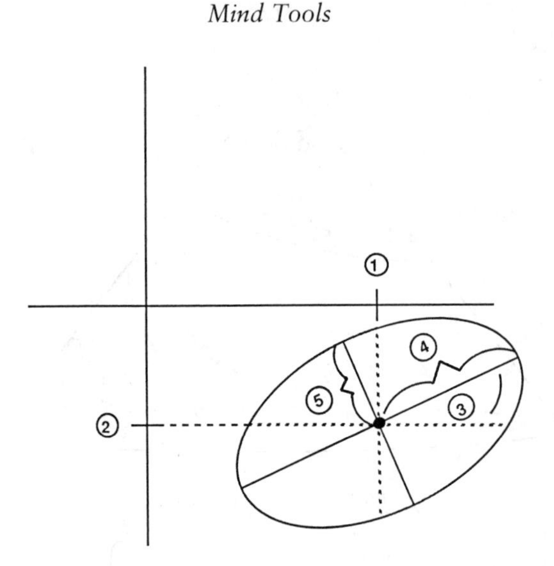

Fig. 56 Five pieces of information specify a general ellipse.

dimensional parameter space based on scale. A flat parabola can be thought of as simply a magnification of a narrow parabola. Put differently, if we move any parabola to the origin and point it upward, then it will have an equation $y = Kx^2$. Only one parameter, the scale parameter K, is free to vary. The larger K is, the faster the y values grow, so the narrower the parabola is.

Although the second-degree algebraic curves make up a five-parameter family, only two parameters are essential to shape; the other three have to do only with position and orientation. It is possible to fit all the possible second-degree algebraic curve shapes into a single two-dimensional parameter space. Going back to the fact that all these curves are gotten by intersecting a plane with a cone, we can take our two parameters to be D, the distance down the cone to the highest point where the plane cuts it, and Θ, the angle the cutting plane makes with the horizontal. Although I didn't mention it above, the cone is taken to be a *double cone,* so that one gets both branches of a hyperbola.

Now we can set up a D–Θ parameter space and get all the conics. The one-parameter family of circles lies on the line $\Theta = 0°$, and the

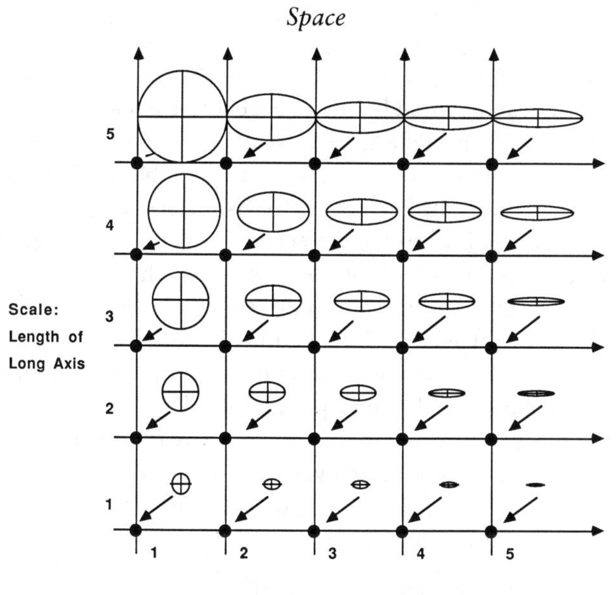

Fig. 57 The ellipse's parameter space.

one-parameter family of parabolas lies on the line $\Theta = 45°$. The ellipses are between the circles and the parabolas, and the hyperbolas are to the right of the parabolas. If we continued the diagram out past $\Theta = 90°$, we'd get the same hyperbolas, then the parabolas again (at $\Theta = 135°$), then the ellipses and the circles.

The conics that lie along the line $D = 0$ are known as "degenerate

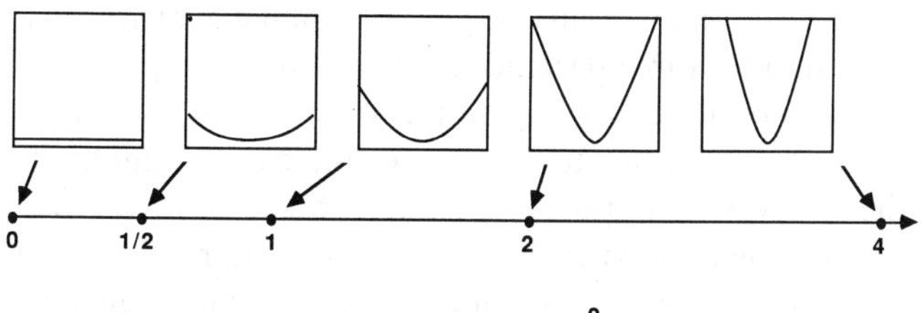

Fig. 58 The parabola's parameter space.

conics." If the cutting plane hits the cone only at the vertex, we get a point, which can be thought of as a degenerate circle or a degenerate ellipse. If the cutting plane goes through the vertex at a 45-degree angle, then it is going to graze along the double cone in a single straight line, which can be thought of as a degenerate parabola. If the cutting plane goes through the vertex at an angle between 45 and 90 degrees, we get a pair of intersecting lines, which can be thought of as a degenerate hyperbola. This mathematical meaning of the word "degeneracy" is not wholly unrelated to the usual sociological sense of the word: isn't someone who stays drunk all the time a bit like an ellipse that lets itself collapse to point size?

What about curves whose equations use powers greater than two? Here things get very complicated very fast. Let's begin by just looking at polynomials — curves whose equations have y on one side and whole powers of x on the other:

$y = x^3 - 12x$ is a cubic polynomial;
$y = x^4 - 8x^2 + 2x$ is a quartic polynomial.

By moving the origin of our coordinate system around, we can turn any cubic polynomial equation into an equation of the form $y = K(x^3 + Ux)$, where K and U are constants. K is what is known as a "normal" parameter — it controls the curve's scale, but not its overall appearance. U is a much more important constant, known as a "splitting" parameter — different values of U lead to wholly different curve shapes. To be precise: If U is negative, then the cubic has a kind of S-shape; if U is zero, then the cubic momentarily flattens out at the origin; if U is positive, then the cubic zooms up with only a ripple.

Thus we see that the shapes of the polynomial cubics make up a two-parameter family of curves. The polynomial quartics, on the other hand, make up a three-parameter family. By moving the origin, any quartic polynomial can be put in the form $y = K(x^4 + Ux^2 + Vx)$. As before, the K parameter just varies the curve's amplitude; the U and the V are the splitting parameters. Since it is hard to draw a three-dimensional parameter space, let us ignore K and draw the parameter space for the polynomial quartic curves in terms of U and V.

This way of drawing the parameter space of the polynomial quartics

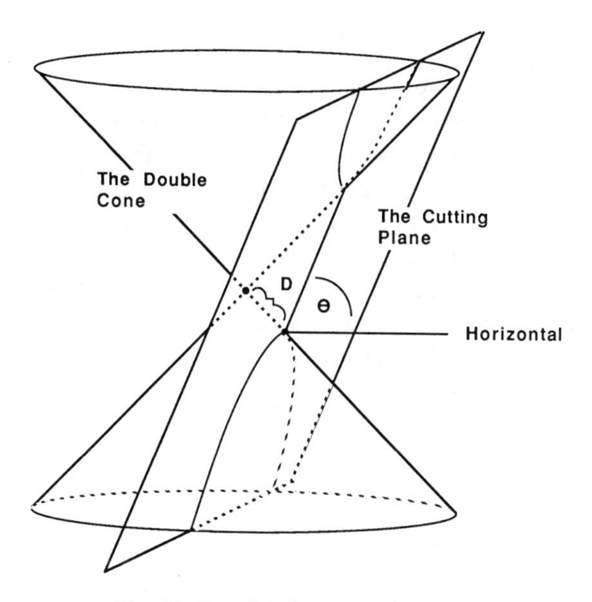

Fig. 59 D and Θ determine the cut.

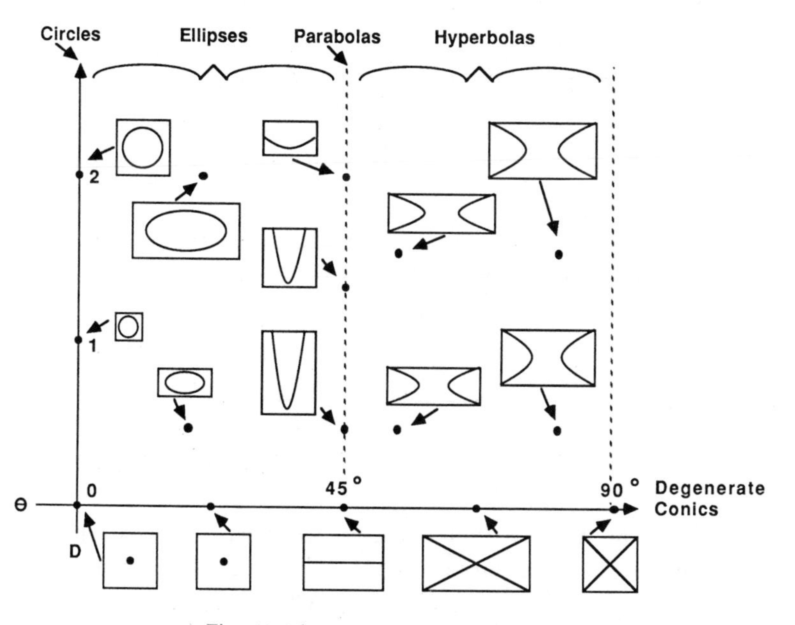

Fig. 60 The parameter space for conics.

happens to be of some significance in the modern mathematical science known as "catastrophe theory." (The postmodern "chaos theory" grows out of catastrophe theory.) Catastrophe theory talks about the various ways in which things can change abruptly. Examples of catastrophes range from things as harmless as a light switch being clicked on to things as violent as earthquakes and prison riots. The significance of the polynomial quartic curves here is, if you think of the curve as being a metal band with a marble resting on it, some smooth changes in the control parameters U and V can lead to an abrupt and jump-like change in the location of the marble.

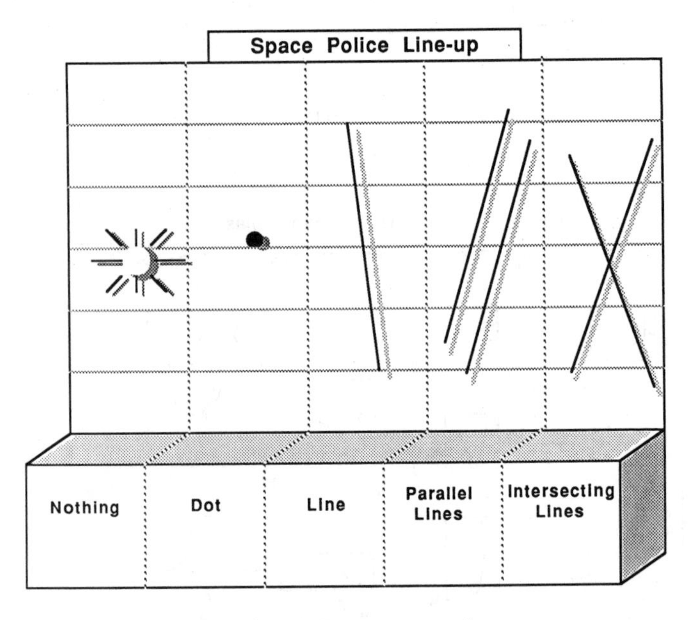

Fig. 61 Degenerate conics.

At this point we have found parameter spaces to represent all second-degree algebraic curves, all third-degree polynomial curves, and all fourth-degree polynomial curves. What about the general cubics — the curves with third-degree equations that cannot be solved for y and put in polynomial form?

A general cubic curve has an equation of the form

$$Ax^3 + Bx^2 + Cxy^2 + Dy^3 + Ex^2 + Fxy + Gy^2 + Hx + Jy = 1 \text{ (or 0).}$$

This means that a cubic is basically given by nine pieces of information: the values of the nine coefficients A through J. If we are only concerned with the shape of the cubic we can ignore four of the nine parameters, as two parameters will determine position, one will determine orientation, and one will determine size. This still leaves five parameters. Figuring out the structure of the five-dimensional parameter space of all possible cubic shapes is no easy task.

In 1676, Isaac Newton came up with the first extensive classification of the third-degree curves. He thought there were seventy-two

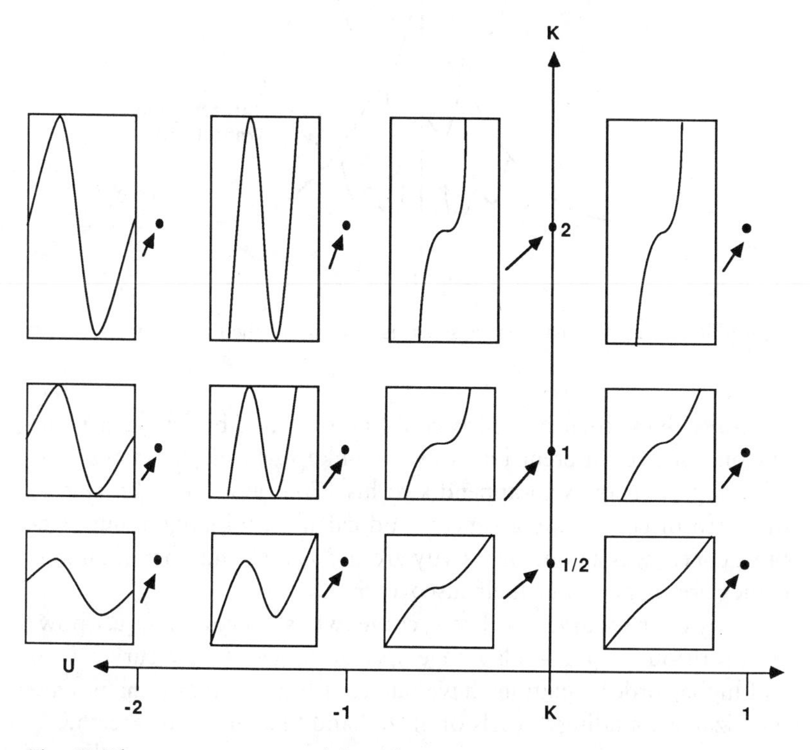

Fig. 62 The parameter space of the polynomial cubics. K and U are values in the equation $y = K(x^3 + Ux)$.

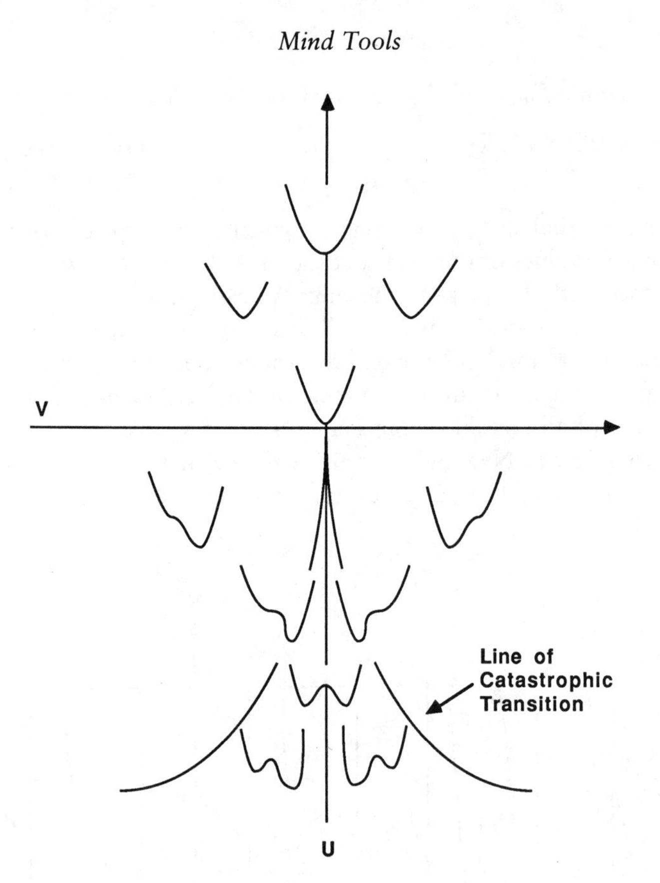

Fig. 63 Polynomial quartic space as catastrophe theorists view it.

general cubics, each of which could be obtained by letting a plane cut one of five different kinds of cone-like surfaces. Newton's classification appears as an appendix to his 1704 book *Opticks*. Due to the birth of coordinate geometry and calculus, thinking about algebraic curves was a popular activity around 1700. A number of famous cubics are named after their discoverers.

Many quartic curves — that is, curves whose *xy* equations use powers up through four — also have special names. A few curves with still higher order equations have names. My two favorite names are the bizarre-sounding "Pearls of Sluze" and "Nephroid of Freeth." It is easy to imagine a science-fiction adventurer who tries to steal some

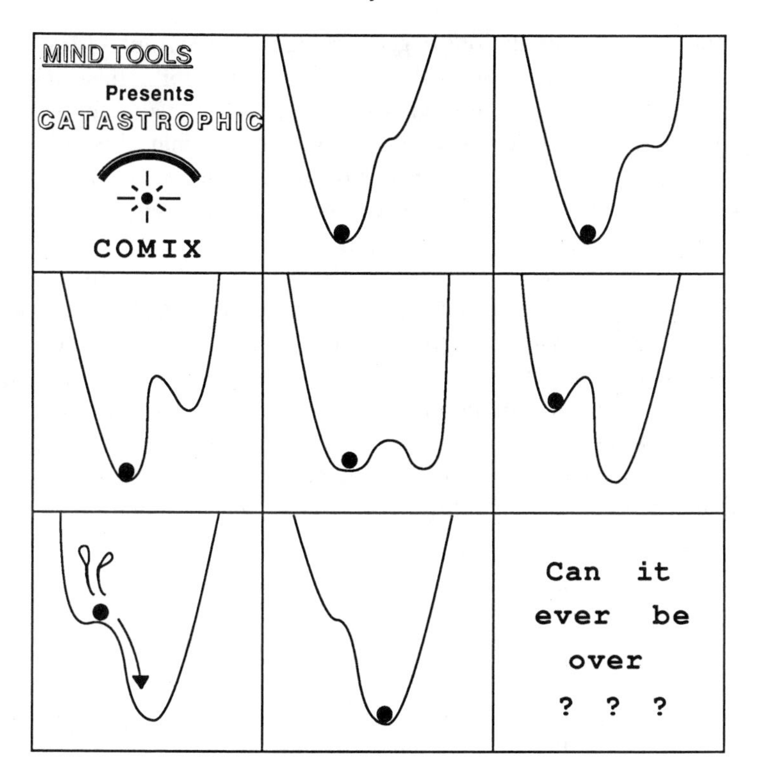

Fig. 64 An abrupt transition.

rubbery and aphrodisiacal Pearls of Sluze from the cruel and deformed Nephroid of Freeth. (In reality, there was a Baron René Française de Sluze sketching his pearls in the 1650s. He was a friend of the philosopher–mathematician Blaise Pascal. The historical identity of "Freeth" has so far eluded me.)

How did those old-time mathematicians manage to graph such complicated curves anyway? For one thing, they took their time. If the only thing you're going to be remembered for in 300 years is one single curve, then you don't mind spending a year or two in getting it right! More seriously, most of the special curves were sketched by using tools from differential calculus.

Given the *xy* equation for an algebraic curve, one can carry out certain simple manipulations to get some related xy formulas for what are known as the *derivatives*. Now, given any value for x, the original equation enables you to estimate the corresponding value for y; the formula for the first derivative enables you to tell what direction the curve is going when it passes through the point with coordinates (x,y); the formula for the second derivative enables you to say which way the curve is bending when it goes through (x,y).

Finding the values of (x,y) that make the various derivatives zero is a good way to locate a curve's most important points. Two simple examples of this involve locating maxima and points of inflection. A curve hits a maximum at any point where the y value is larger than it is at any other nearby point on the curve. A curve hits a point of inflection when it goes from being cupped one way to being cupped the other way. The purpose of looking at derivatives in this situation is that whenever the curve hits a maximum (or a minimum, for that

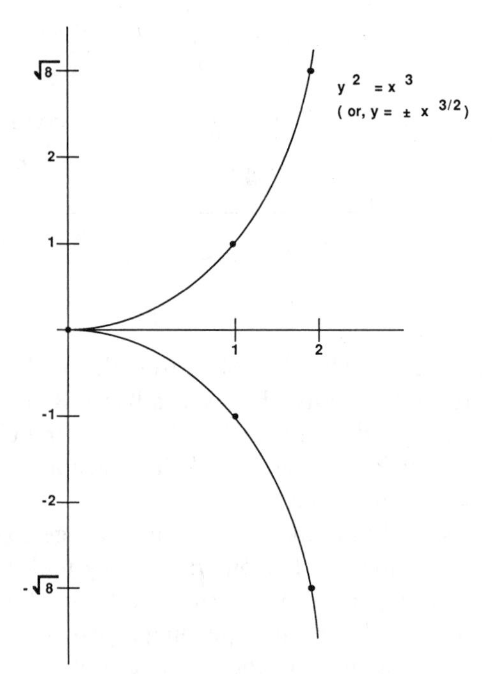

Fig. 65 The semicubical parabola is the graph of the simplest third-degree equation for which y can't be written in terms of whole powers of x.

matter), its first derivative is zero; whenever a curve hits a point of inflection, its second derivative is zero. Other kinds of derivatives give other information.

Of course, nowadays one can draw very complex curves even if one knows nothing at all about calculus and the craft of curve sketch-

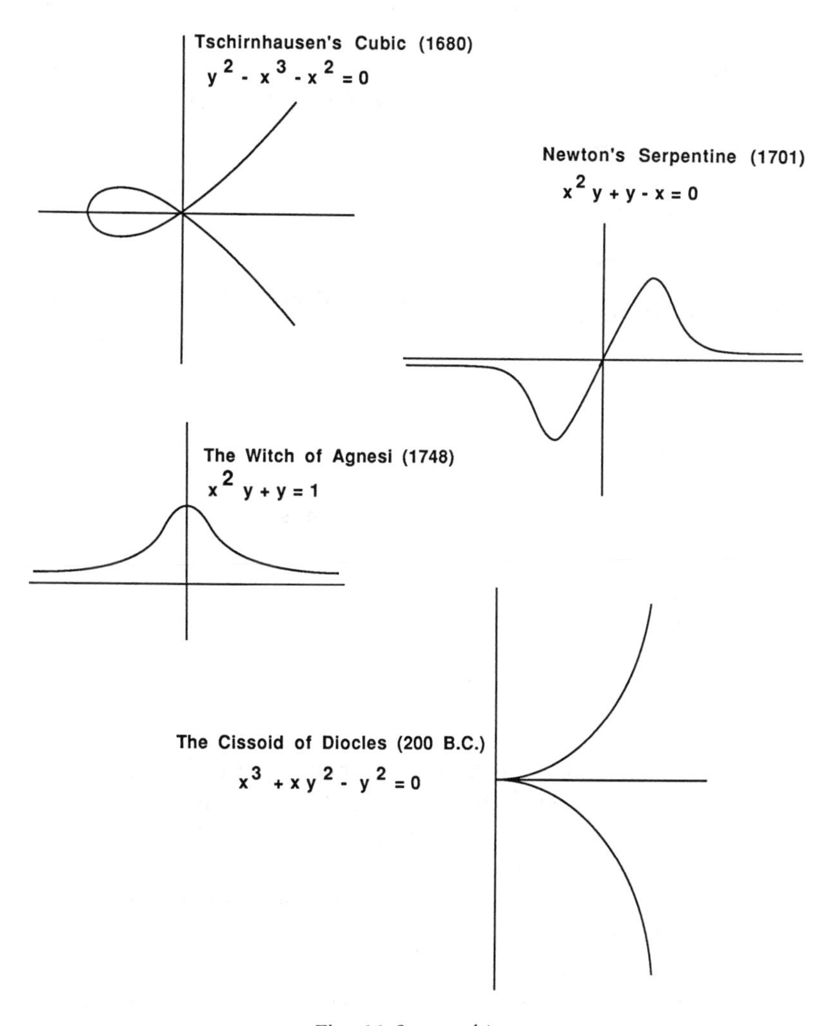

Fig. 66 Some cubics.

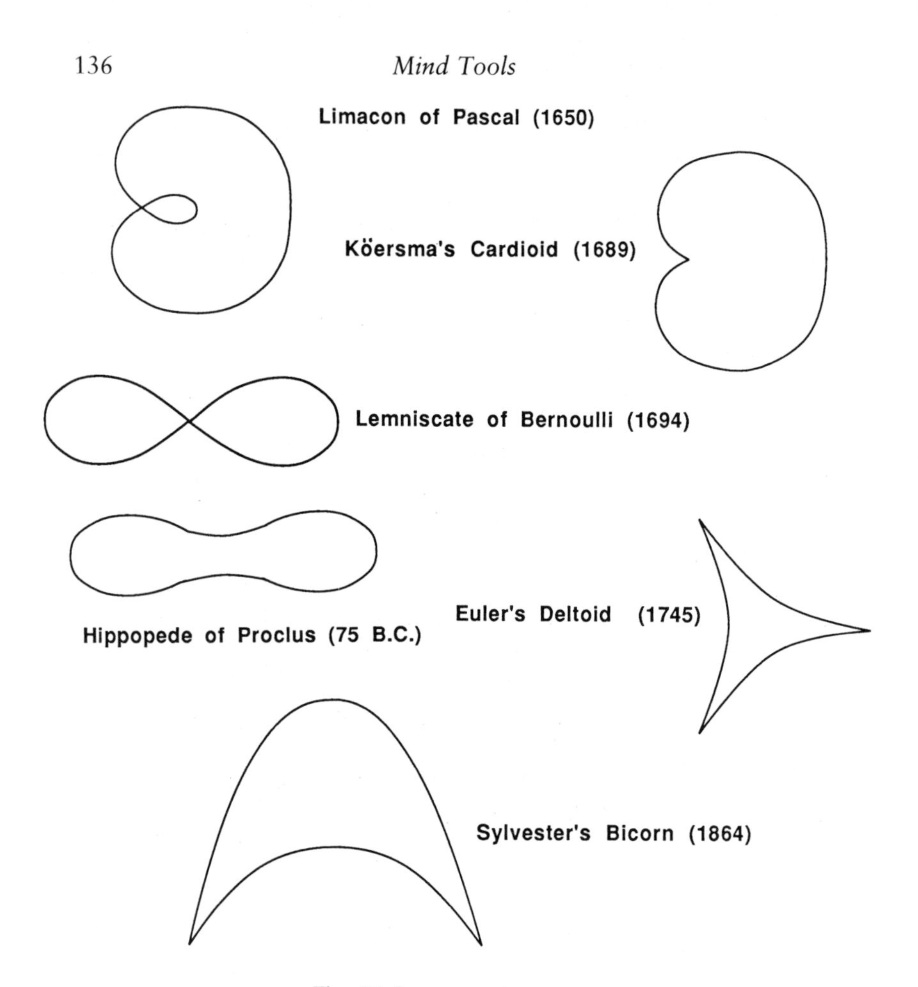

Fig. 67 Some quartic curves.

ing. It is only necessary to own a computer equipped with some simple graphics software; even LOGO will do. Given the equation of an algebraic curve, one makes a rough estimate of how big the curve's interesting part is and sets up an appropriately scaled coordinate system on the screen. Then one runs a series of brute-force calculations to generate a few hundred bright dots located at points on or near the curve. If you stand back from the screen a bit, the dots merge into what looks like a smooth, bright line. The curve is sketched. If you need a permanent record, you can photograph the screen or just make a hard copy on a printer.

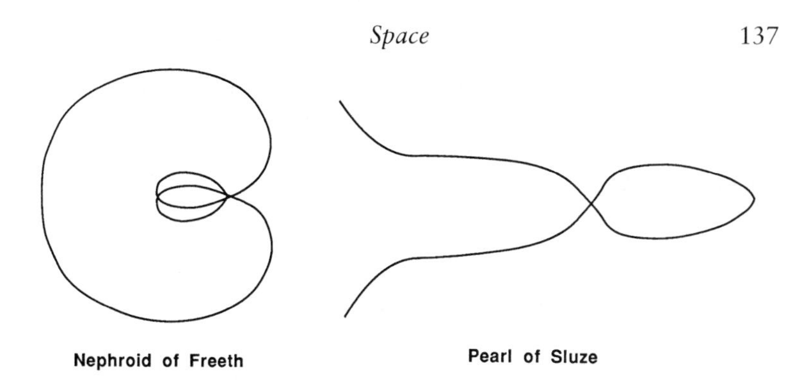

Fig. 68 Two high-degree algebraic curves with funny names.

What I find most remarkable about the algebraic curves is the complexity of form that arises out of so few parameters. Looking at some weird, warped curve, the natural tendency is to think of the curve as infinitely complex, yet if the curve is algebraic, it is determined by just a few real-number parameters — parameters that we think of as individual bits of analog information.

Recall that, within reasonable precision, an analog position bit corresponds to about seven bits of digital information. So a few-score digital bits can determine curves of high degrees. It is constantly surprising to find out how complex-looking a pattern can be gotten out of a few information bits. Stringing together analog bits is like specifying more and more parameters. When I say that the family of elliptical shapes makes up a two-parameter family, I am really saying that, viewed as analog information, an ellipse's information content is two analog bits, which again, at the seven-to-one conversion rate, is about 14 digital bits.

Make a quick, random scrawl on a piece of paper. How much information does this scrawl code up? If I reflect on the actual process of doing a scribble, it seems like the process involves making about ten consciously random analog choices. The geometry of my arm and hand probably input another ten parameters or so. Does this mean there is an algebraic curve of, say, degree fifteen that matches my squiggle? Theoretically there are infinitely many squiggles I might have made, but in practice must I not light on a standard curve in some low-dimensional parameter space?

Let's simplify the problem. Given a bunch of points in the plane, can I find an equation for an algebraic curve that passes through each of these points? Yes. This process is sometimes called "curve-fitting." One crude way to curve-fit is as follows: Suppose I have targeted K points with coordinates (a_1,b_1), (a_2,b_2), . . . (a_K,b_K). Now write the xy equation

$$(x + y - a_1 - b_1)(x + y - a_2 - b_2) \cdots (x + y - a_K - b_K) = 0.$$

The equation will be of degree K, and it will be satisfied by each of the K original points. Therefore the graph of this equation will be an algebraic curve going through each of my original K points. This particular curve is determined by $2K$ analog bits of information. The more precisely I specify the a and b coordinates, the more precisely the curve will be given.

Now go back to the random squiggle. Given a squiggle, we can set up a coordinate system, pick a reasonably dense bunch of points on the curve (fifty seems like enough points for my particular squiggle), measure the coordinates of the points (to some reasonable level of accuracy), and use a simple curve-fitting procedure to produce an

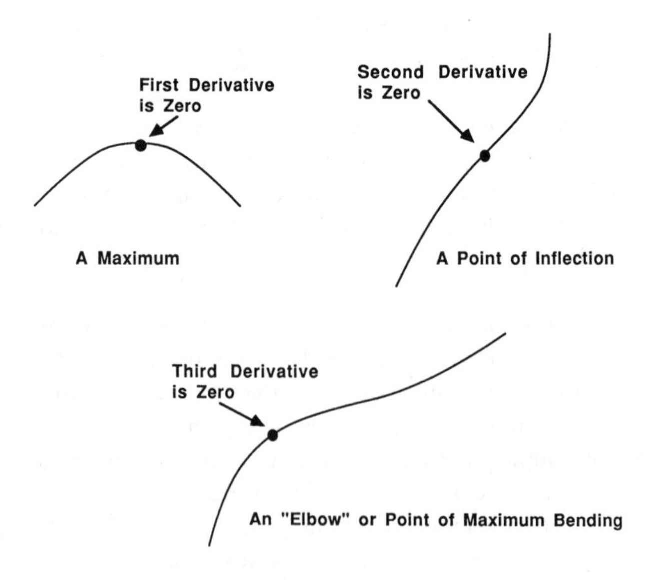

Fig. 69 Zero derivatives mark points of interest.

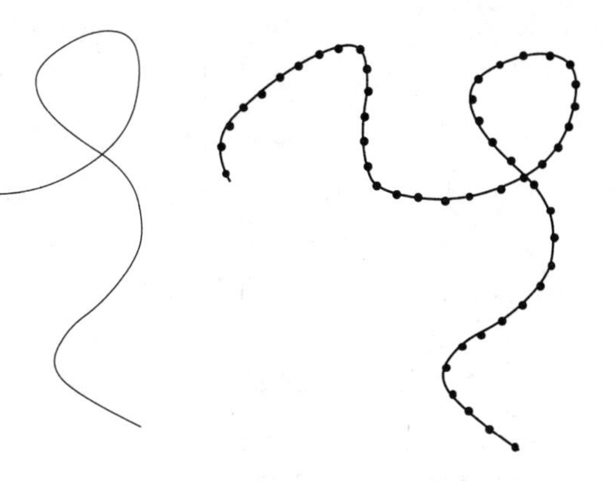

Fig. 70 A scrawl. **Fig. 71** One hundred analog bits.

equation for a curve passing through the points in question. Although this fact is not obvious, it turns out that our procedure can be adjusted to give a curve that connects the dots as smoothly and directly as possible, without any extra jumping around.

Actually figuring out the equation of my squiggle would take a lot of calculation, but just like the sketching of curves, this process can be computerized. Computers like the Macintosh and the Amiga have programs that take as input a squiggle drawn on the screen with a table mouse. It would be feasible to write a program that would yield as output the equation of an algebraic curve that approximates the squiggle to any degree of accuracy up to the limits determined by the screen's grid or pixel size.

Given the machinery of algebraic curves, we see that a random, messy squiggle involves only some one hundred analog bits. Let's see exactly how effective this machinery is by comparing it to some other ways of describing squiggles. In order to have a definite standard, let's assume that each analog bit costs seven digital bits. Then approximating a squiggle by finding an algebraic curve through fifty of its points means using one hundred analog positions, which is seven hundred digital bits.

In speaking of an analog bit as worth seven digital bits, I am really doing the following. I am thinking of my paper as divided into a pixel mesh measuring 128 by 128 cells — 128 is two to the seventh, so it

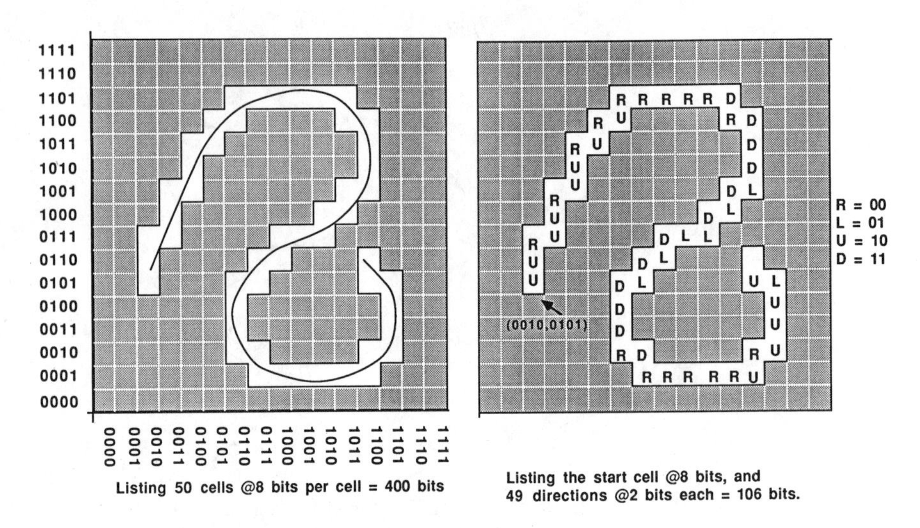

Fig. 72 Two ways of coding a curve.

takes seven bits to specify a horizontal or vertical position, fourteen bits to specify an individual cell.

Suppose I take a crude approach and describe my squiggle by listing the cells it passes through. An arbitrary squiggle won't hit most cells, so this method isn't too inefficient. Assuming the squiggle twists around a moderate amount, it might go through five hundred different cells. If we list each of these cells at a cost of fourteen bits a cell, we need seven thousand bits. This is ten times as costly as finding the smoothest algebraic curve to connect fifty of the curve's cells. Of course, if the curve was very spiky, then the latter method might give a better model.

An approach intermediate in subtlety is to take advantage of the fact that the curve is continuous, but without bringing in much algebra. Since the squiggle is continuous, if it passes through a cell, then it must also pass through one of the cell's neighbors. Let's assume that a squiggle must leave a cell through one of the cell's four sides: *R*ight, *L*eft, *U*p or *D*own. Now a choice among four possibilities only costs two bits, so once we have the fourteen bits that give the curve's starting point, we can specify the 499 next cells at a cost of only two bits a cell, which means a total cost of 1012 bits in all.

This last method is probably the one that a computer would be most likely to use in describing a curve. It takes a bit more memory than the algebraic curve approach, but it involves a lot less calculation.

Wiggles and Whorls

As I mentioned in the first section of this chapter, it is sometimes said that there are six kinds of simple natural curves: polygons, circles, waves, helices, spirals, and branches.

The polygons and their space-cousins the polyhedra were touched on in the chapter on number and again in the section on tessellation. The circle is a special sort of algebraic curve. Many algebraic curves besides the circle also arise in nature — a ball thrown upward travels in a parabolic path; a sausage cut at an angle yields elliptical slices; a one-shot comet traces a hyperbolic path through space.

The vast majority of higher-degree algebraic curves are seen more rarely, although there are many exceptions such as the curves in the family of "caustics." Caustics are bright patterns of light created when light is bounced off or refracted through any kind of curved surface, such as a mirror or a lens. When light bunches up to make bright lines, these lines are called caustic lines because "caustic" comes from the Latin for burning. If you try to burn something with a magnifying glass, the bright lines around the hot spot are caustics. The loosely

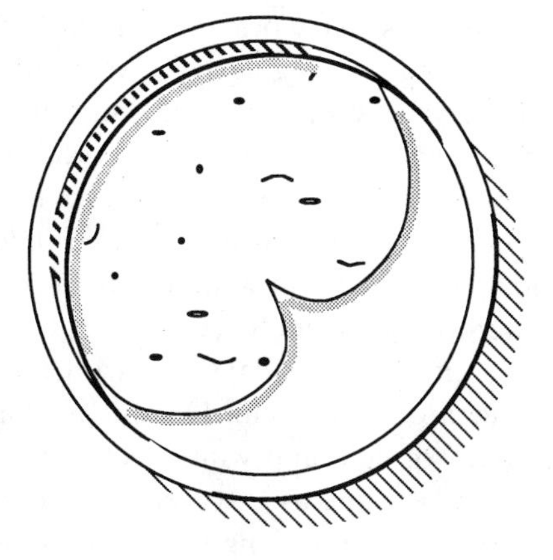

Fig. 73 Caustic nephroid in a wedding ring.

woven light patterns you see on the bottom of a swimming pool are caustics. The simplest caustic curve is the nephroid (an algebraic curve of degree six). You can see it on the surface of your coffee, provided that your cup is shiny and sitting in the sun; alternatively, if you have a wedding ring, set the ring down on a sunny windowsill to see the nice curve of degree six.

But what about the other kinds of basic curves — the waves, spirals, helices, and branchings? These naturally arising curves are nonalgebraic for different kinds of reasons. An endless wave fails to be algebraic because no algebraic curve has an endless number of wiggles. Spiral patterns do not readily lend themselves to description in terms of the rectangular coordinates of cartesian geometry. A deeper reason for which the spiral pattern of a seashell is nonalgebraic is that the shell arises from a process of exponential growth. Helical, or corkscrew, patterns are nonalgebraic because, like waves, they oscillate back and forth endlessly. Another difficulty with helices is, of course, that these curves twist out of the plane and into three-dimensional space. We will put off the question of branching curves until the section on fractals. For now, let's focus on wiggles and whorls.

The course of a river characteristically weaves back and forth in a meandering pattern. This pattern can be seen on a small scale in rivulets of water running down a car's windshield. A simple mathematical curve that has the property of zig-zagging back and forth is the *sine curve.*

The sine curve dominates our world more than almost any other shape. Just about anything that regularly waxes and wanes does so according to a sine-curve law of variation. Pure musical tones are sine-curve variations of air pressure with respect to time. Pure colors are sine-curve variations of electromagnetic field strength with respect to time. Large ocean swells are sine waves, as are ripples in a puddle, and as are the moving kinks in an agitated jump-rope. The ocean tides rise and fall according to sine curves keyed to the motions of Earth and moon. The lengths of our days vary with a sine curve keyed to the Earth's journey around the sun. One might think of summer as a kind of "high tide" of light and winter as a thermal "low tide."

The family of sine-wave shapes is a two-parameter family. The two parameters are amplitude and wavelength. The amplitude is a vertical scaling factor that determines the height of the sine curve's waves

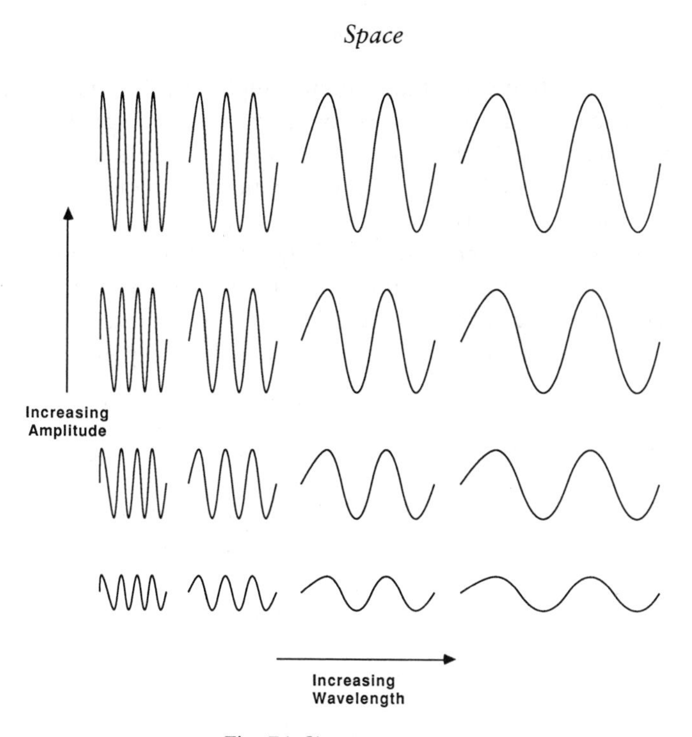

Fig. 74 Sine wave space.

from crest to trough. The wavelength is a horizontal scaling factor that determines the width of the sine curve's waves from crest to crest. The rough wavelengths of, respectively, light waves, ocean waves, and radio waves are one-millionth of a meter, ten meters, and three hundred meters. Are we on the same wavelength?

Instead of wavelength, people often talk about frequency. A sine wave's frequency is inversely proportional to its wavelength; that is, a high-frequency wave has a short wavelength, and a low-frequency wave has a long wavelength. For sound waves, amplitude is experienced as loudness and frequency is experienced as pitch. For light waves, amplitude is experienced as brightness and frequency as hue. The light waves to watch out for are the ones with really high frequency — ultraviolet, x-ray, and gamma ray. Those ultrahigh-frequency waves pack a powerful punch.

AM, or *amplitude-modulated,* radio waves are electromagnetic sine waves with fixed frequency but variable amplitude. The amplitude variations code up the signal. FM, or *frequency-modulated,* radio waves are sine waves with fixed amplitude but variable frequency. Here the signal is coded up by (slight) variations in frequency.

As we mentioned above, no algebraic curve has the sine wave's property of looping up and down infinitely often. The sine cannot be expressed by any algebraic equation of finite degree. For this reason the sine curve is spoken of as "transcendental." It is, however, possible to express the sine wave as the graph of an endless polynomial:

$$y = x - {}^1/_{3!}x^3 + {}^1/_{5!}x^5 - {}^1/_{7!}x^7 + {}^1/_{9!}x^9 - {}^1/_{11!}x^{11} + \cdots$$

In this endless equation, "3!" stands for "1 × 2 × 3," "5!" stands

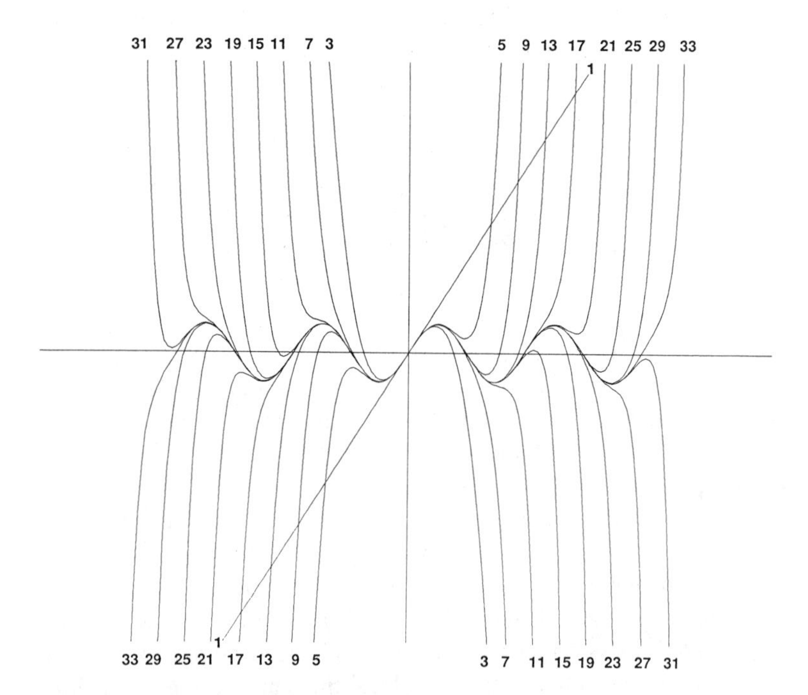

Fig. 75 Polynomial approximations to the sine curve. The curve marked 1 is the graph of $y = x$; the curve marked 3 is the graph of $y = x - {}^1\!/_6x^3$; the curve marked 5 is the graph of $y = x - {}^1\!/_6x^3 + {}^1\!/_{120}x^5$; and so on. This computer graphic was produced by H.R.P. Ferguson and T. Sederberg of Brigham Young University.

for "$1 \times 2 \times 3 \times 4 \times 5$," and so on. The further you go to the right, the smaller the terms get, and the sum of the whole right side is always finite. As one includes more and more terms of this endless polynomial, one gets better and better approximations to the full sine curve.

This is, of course, an unnatural way to generate a sine curve. An endless equation looks like it uses an endless amount of information, and if sine curves needed endless amounts of information, nature wouldn't be so full of them. In reality, a sine curve arises whenever we convert circular or repetitive motion into linear motion. If you put a piece of chalk in your hand and walk along a wall with your arm oscillating up and down, you'll draw a rough sine wave. The repeated arm oscillations are stretched out. There are really only two parameters involved here, two pieces of analog information: the amplitude of your arm's oscillations and the wavelength distance that you walk during one full oscillation.

Another circle plus line instance of the sine wave can be observed if you put a cigarette butt on the edge of an LP record on a moving turntable and squat to put your eye at the record's level. You'll see the butt oscillating from left to right with sine-wave speed; that is, the distance from the spindle is a sine function of the time. The record player example brings up two other important curves: the spiral and the helix. The actual groove on a record is a special kind of spiral known as an "Archimedean spiral," and if one were to lift the record player up through space at a uniform speed, the butt on the edge of the spinning record would trace out a helix whose shadow is, incidentally, a sine wave.

But before we move on to spirals and helices, we should mention another family of transcendental curves just as important as sine curves. I am thinking of exponential curves.

In grade school we learn to add, and then we learn to multiply. In high school we learn the next logical operation, exponentiation. Just as two *times* three means two plus two plus two, two raised to the third means two times two times two. The notion of exponentiation can be extended to fractions and to negative numbers. An exponential curve is a curve with an equation of the form $y = A^x$.

The shape of an exponential curve can be characterized by saying that it starts out looking out like a horizontal line, then passes through

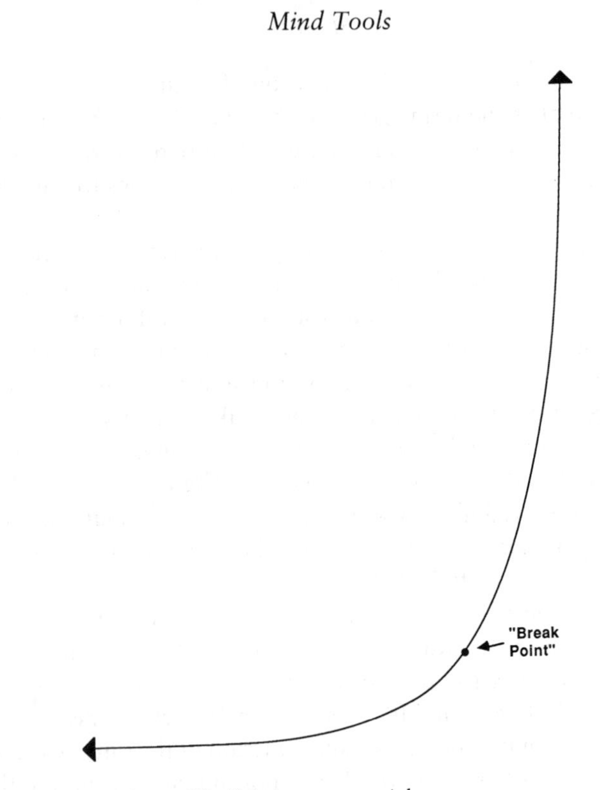

Fig. 76 An exponential curve.

a "break point" where the slope is 45°, and then quickly shoots up faster than any polynomial curve. The property that characterizes the exponential curve is that the higher it gets, the steeper it gets. To say that y is an exponential function of x is to say that y's rate of growth is everywhere proportional to y's size.

People concerned with overpopulation often point out that Earth's population is growing exponentially. The reason population grows exponentially is that the more people there are, the more new babies there are per year. The population's rate of growth is directly proportional to the size of the population. A few centuries ago, this was a good thing, but now we've passed our population curve's "break point" and the population is speeding out toward infinity. Overpopulation can only lead to famine, war, and plague; these bad things are mathematically inevitable if we continue to reproduce at the same rate per person. As we seem unable to change our patterns of reproduction, shouldn't we just resign ourselves to a lot more plagues, wars, and famines in the future? Why get so upset every time a lot of people die? It's really no different from bacteria in a Petri dish. When there are too many bacteria to survive, big patches of them die off and a new growth cycle begins. The bacteria culture will spread right back over the bald spot in a few generations. In the same way, no one is upset if a tenth of the crabs in Chesapeake Bay are steamed to death every year. One crab is much like another, and our depredations leave the race of crab not one whit diminished. Why don't we feel this way about famine in Ethiopia, typhoons in Bangladesh, or an earthquake in Mexico? Most people are about the same, so there's no great global information loss if a few million disappear. The missing people's genetic information will be reproduced by their relatives, and the missing people's mental information (skills, experience, knowledge) is undoubtedly shared by others. The argument seems perfecly logical, but if I accept it, I'm put in a position of having to accept the unacceptable statement: "It makes no real difference if all my friends and relatives drop dead tomorrow." The difference for me between people and crabs seems to be that *I* am a person, and I care about people as individuals. Perhaps a crab sees things differently.

Like the sine curve, the exponential curve is transcendental — i.e., not expressible by any finite algebraic formula (only positive numbers are allowed as exponents). The reason is that the exponential curve rises more steeply than any polynomial function. Again like the sine curve, the exponential curve can be expressed as the graph of an endless polynomial:

$$y = 1 + x + {}^{1}/_{2!}x^2 + {}^{1}/_{3!}x^3 + {}^{1}/_{4!}x^4 + {}^{1}/_{5!}x^5 + \cdots.$$

It is worthwhile reminding the reader here that it is meaningful to look at negative exponents: In general, $A^{-x} = 1/(A^x)$. Thus ten to the negative one is one-tenth, ten to the negative two is one-hundredth, and ten to the negative N is 0.000 . . . 0001, where there are N places to the right of the point. By the same token, two to the negative one is one-half, two to the negative two is one-fourth, and two to the negative N is the binary point fraction $0.000 \ldots 0001_{\mathrm{TWO}}$, where there are N places to the right of the point.

One last important transcendental function is the logarithm func-

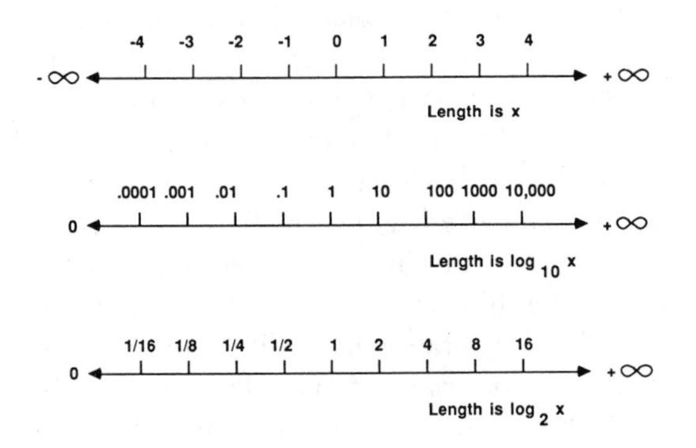

Fig. 77 A number line and two number-log lines.

tion. As mentioned in chapter 1, the logarithm can be defined as the inverse of the exponential. That is, we say $y = \log_A x$ means that $A^y = x$. Our remarks about negative exponents mean that numbers between zero and one have negative logarithms:

$$\begin{aligned} \log_{10} 0.1 &= -1; & \log_2 {}^1\!/_2 &= -1; \\ \log_{10} 0.01 &= -2; & \log_2 {}^1\!/_4 &= -2; \\ \log_{10} 1/\text{million} &= -6; & \log_2 {}^1\!/_{1024} &= -10. \end{aligned}$$

It is often useful to draw number-log lines; that is, number lines on which each number is represented by a length corresponding to its logarithm rather than its size. Note in Fig. 77 that the whole left half of the number-log line is filled with numbers between zero and one; the logarithm of a negative number is not defined. Put a bit differently, the log of zero is negative infinity.

In chapter 1 we noted briefly that if we use a number-log line for the x axis and an ordinary number line for the y axis, the logarithm function graphs as a straight line. Conversely, if we use an ordinary number line for the x axis and a number-log line for the y axis (Fig. 78), exponential functions take the form of straight lines. If we use "log log" graph paper, on which both axes are number-log lines (Fig. 79), then an equation of the form $y = Kx^N$ graphs as a straight line with slope N.

Let's look at some spirals (Fig. 80). The Archimedean spiral, which we mentioned above, is characterized by uniform spacing between its

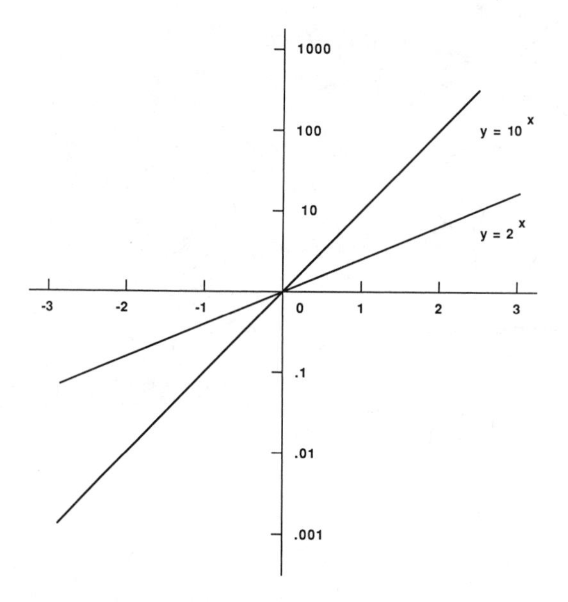

Fig. 78 Graphs of exponential functions on number-line versus number-log-line paper.

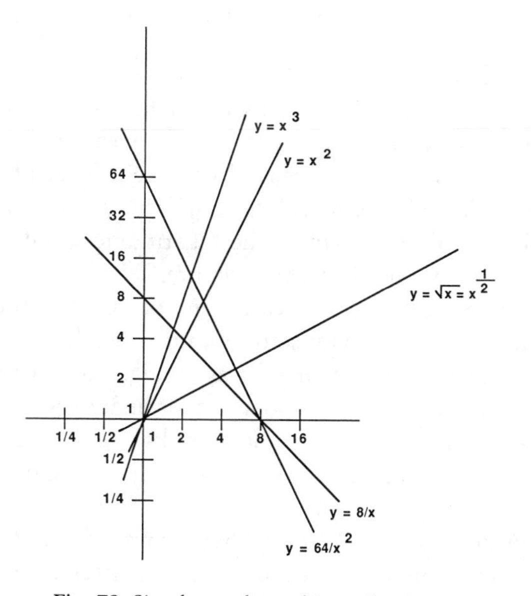

Fig. 79 Simple graphs on binary log log paper.

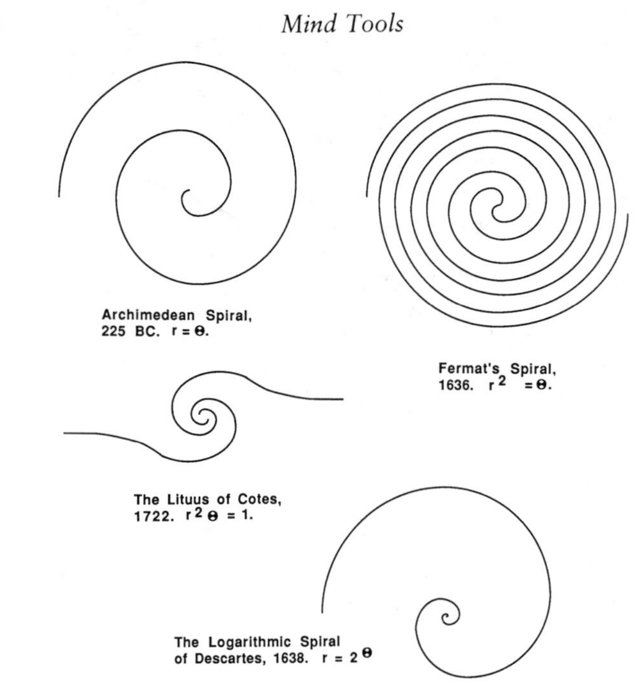

Fig. 80 Some spirals.

successive arcs. It's about as exciting as a straight line. A hotter spiral is Fermat's spiral, which has the hallucinatory appearance of an endless yin–yang. The Lituus of Cotes is nice; if we just take half of it, we get something resembling the volute of an Ionic column. The most important spiral of all is known as the logarithmic spiral; it is important because it is found so often in nature.

Until now we have been using cartesian coordinates to describe curves, but the way a spiral wraps around the origin infinitely many times makes it impossible to express it as an algebraic xy equation. For spirals it is better to use a "polar" coordinate system. As the name suggests, a polar coordinate system looks like the mesh of latitude and longitude lines that one sees at the poles of a globe map of the Earth.

In polar coordinates, all our spirals have simple descriptions. We can imagine tracing an Archimedean spiral, for instance, by taking the path of a point that moves around and around the origin at a

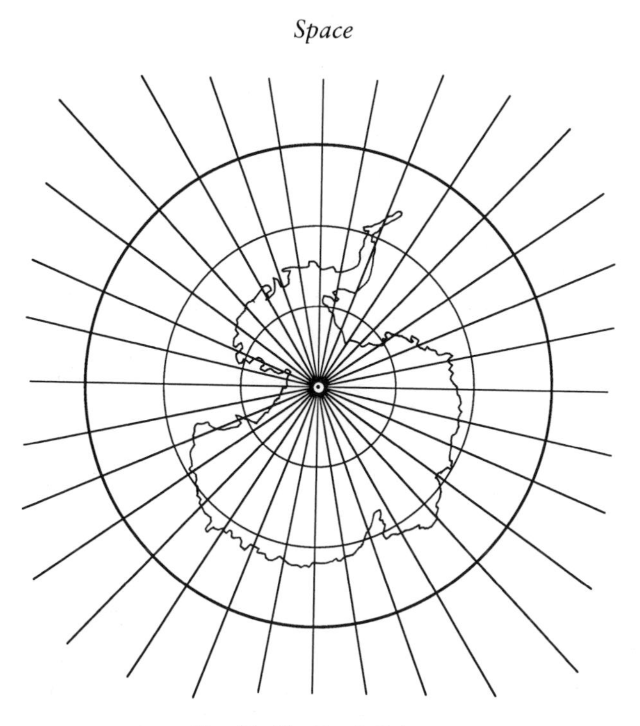

Fig. 81 The South Pole.

constant rate while the point's distance from the origin increases at a constant speed. What this means, if you think about it, is that the tracing point's distance r from the origin is directly proportional to the angle Θ that measures how many times the point has moved around the origin. In other words, the polar-coordinate equation for an Archimedean spiral has the simple linear form $r = A \cdot \Theta$, and the equation of the logarithmic spiral has the form $r = A^{\Theta}$, for some constant A. Really, the logarithmic spiral might just as well be known as the exponential spiral, though if you want to get a logarithm involved, you can write its equation as $\Theta = k \log r$.

The characteristic feature of the logarithmic spiral is that corresponding pieces of this spiral are always the same shape as each other, in the mathematical sense of having their component parts bear the same constant ratios to each other. The importance of mathematically similar figures is that they look like enlarged or shrunken versions of each other. As an example of this property of the logarithmic spiral,

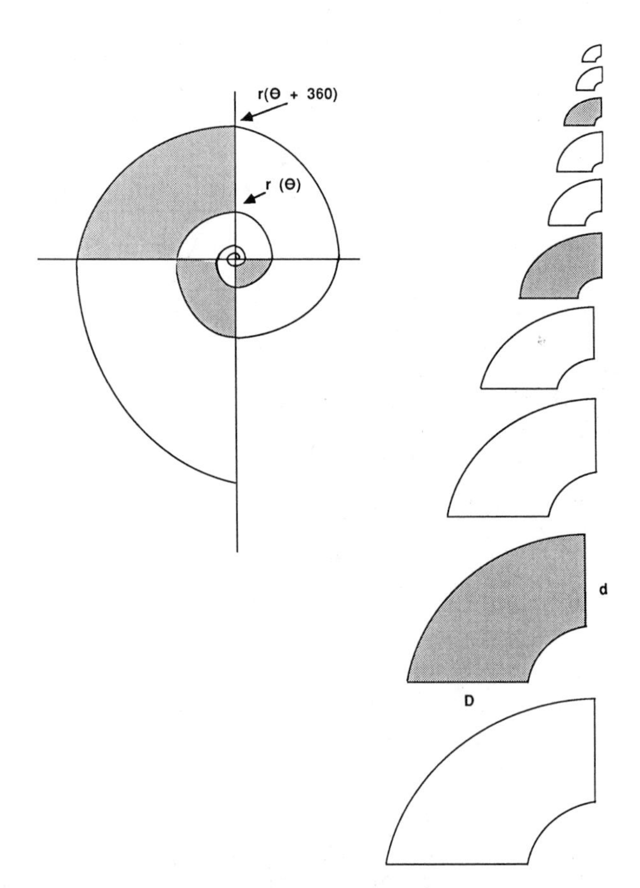

Fig. 82 Corresponding parts of the equiangular spiral have the same shape.

we now prove that if the spiral in Fig. 82 has the equation $r = A^{\Theta}$, then the indicated sectors are all similar.

To be more precise, we want to show that in every ninety-degree whorl sector the long straight edge D has the same ratio K to the short straight edge d. Note that the spiral comes back on itself every 360 degrees, so the distance across a whorl is going to have the form $r(\Theta + 360) - r(\Theta)$, for some value of Θ. Now,

$$d = A^{\Theta + 360} - A^{\Theta} = A^{\Theta} \cdot [A^{360} - 1]$$
$$D = A^{\Theta + 90 + 360} - A^{\Theta + 90} = A^{\Theta + 90}[A^{360} - 1] = A^{90}d.$$

Therefore $D/d = A^{90}$, which is a constant K with the same value everywhere along the curve.

So a logarithmic spiral has the property that corresponding parts of it have the same shape. The converse can also be proved; that is, it is possible to show that a spiral curve with the property that corresponding parts have the same shape *must* be a logarithmic spiral. In his monumental classic, *On Growth and Form,* D'Arcy Thompson explains that is why the logarithmic spiral — which he also calls "the equiangular spiral" — appears over and over in nature:

> It is characteristic of the growth of the horn, of the shell, and of all other organic forms in which a [logarithmic or] equiangular spiral can be recognised, that *each successive increment of growth is similar, and similarly magnified, and similarly situated to its predecessor.* . . .
>
> We must never expect to find the logarithmic spiral manifested in a structure whose parts are simultaneously produced, as for instance in the margin of a leaf, or among the many curves that make the contour of a fish. But we most look for it wherever the organism retains, and still presents at a single view, the successive phases of preceding growth: the successive outlines occupied, as growth pursued the even tenor of its way. And it follows from this that it is in the hard parts of organisms, and not the soft, fleshy actively growing parts, that this spiral is commonly and characteristically found: not in the fresh mobile tissue whose form is constrained merely by the active forces of the moment; but in things like shell and tusk, and horn and claw, visibly composed of parts successively and permanently laid down.

Thompson is saying that, *because* molluscs grow their shells so that equal-sized whorl sectors are similar to each other, the shells look like logarithmic spirals. The logical connection between the first clause and the second clause can be demonstrated mathematically, and the first clause holds because of the way a snail grows its shell.

Consider: A young snail looks exactly like an old snail, except that the old snail is larger. Now, the successive increments to the snail's shell are always being produced by calcium-secreting patches on its slug-like body. These patches grow with the body, but they retain their shape. The consequence of this is that the shell, which is like a series of discarded suits, curves into a self-similar spiral. The same reasoning holds not only for shells, but for horns, teeth, claws, and tusks. An elephant's tusk grows out of a living tusk bud in its skull. As the elephant grows, the tusk bud grows, too, but without changing its shape. For that matter, even the outer surfaces of the teeth in your

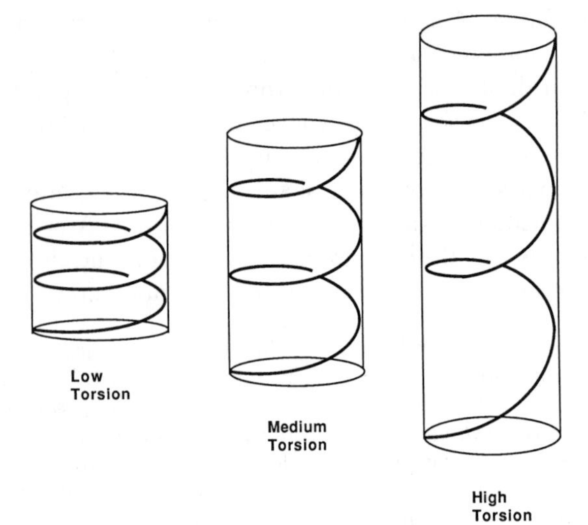

Fig. 83 Twisted space curves.

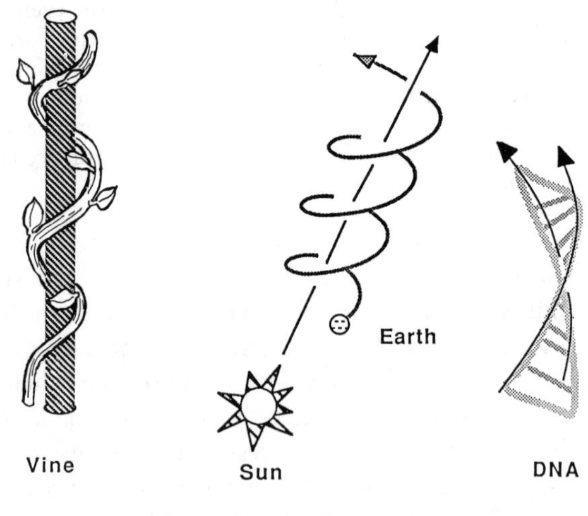

Fig. 84 Natural helices.

mouth are segments of equiangular spirals. Perhaps if you could see back through time, your many past bodies would curve out from you in some uncanny logarithmic spiral of four dimensions.

If we put a straight line out in space, it is still just a straight line. The simplest curves that really need three dimensions to exist are the spirals' cousins, the helices. A helix is a curve that twists around and around a cylinder. Corkscrews, spiral staircases, toy slinkies and light-bulb filaments are all helices. As well as having curvature — the usual bending of a plane curve — a helix has what is known as "torsion." Torsion measures a space curve's tendency to twist out of the plane.

All kinds of climbing vines naturally grow in helical patterns. As our sun moves along its own line of motion, our planet's circling about the sun traces out a kind of helix. Our genetic information is coded up by DNA molecules, which have the shape of twin helices.

On a metaphorical level helices are very interesting, since they combine circular motion with linear upward motion. Human intellectual development can be thought of as helical. Society is always circling back and forth between liberalism and conservatism, yet the conservatism of the 1980s is not quite the same as the conservatism of the 1950s. Yesterday's cycle through liberalism informs today's conservatism, just as today's conservatism will inform tomorrow's liberalism. The motion is helical, rather than circular. On one level we might say that we are back where we started, but this is not the case. The past is not forgotten.

The same kind of helical pattern can be seen in the traditional quest for spiritual enlightenment. A person starts out leading an ordinary life, but then begins to long for some kind of higher truth. Setting out on a quest for deeper meaning leads, say, to meditation and various other spiritual exercises. At the end of such a spiritual quest, one may end up by deciding to lead an ordinary life. This is reasonable. The "secret of life" is life itself. On the surface, this appears circular, but the seeker has really taken a loop around a helix. The very fact of having made the quest changes the nature of the "ordinary life" to which one returns. (See Fig. 85.)

Endless Complexity

Mathematics tries to replace reality with a dream of order. It is perhaps for this reason that mathematicians are often such strange and socially inept people. To devote oneself to mathematics is to turn away from the physical world and meditate about an ideal world of thoughts. The striking thing is that these pure mathematical meditations can in fact make fairly good predictions about messy matter. Eclipses are predicted; bridges are built; computers function; blah, blah, blah.

Mathematics proceeds by extracting simple forms from the world, although later it is often desirable to complicate these forms enough to be good models of reality. In the early stages of geometry, one talks only about such simple shapes as spheres and cubes. Roughly speaking, the Earth is a sphere. The shape of a room is more or less that of an elongated cube. The horizon is approximately a straight line. Viewed from a distance, a parking lot is much like a section of a plane. It is sometimes enough to model the world, in its complexity, by such simple mathematical forms.

The good thing about simple forms is that it is easy to reason about them. Computing the volume of an elongated cube is simple (length times height times width), whereas computing the volume of a real room would be quite hard, if you keep in mind that the corners aren't going to be perfect right angles, that the line where wall meets ceiling is not quite straight, that the bumpy walls are not perfectly plane, and that physical space is a bit "curved" and thus not truly Euclidean.

Simple models are useful, but in the long run, mathematicians find it too galling to settle for approximations to reality. The ultimate dream of the mathematician is to find a wholly abstract model for reality in all its richness. Whether or not they realize it, most mathematicians subscribe to Plato's notion that the physical world is but a shadow of the pure world of ideas.

If I look at the badly plastered wall of the room in which I'm writing, I notice, first of all, that the wall is scarred by a number of large bumps and gouges. If I get up and examine one of the bumps

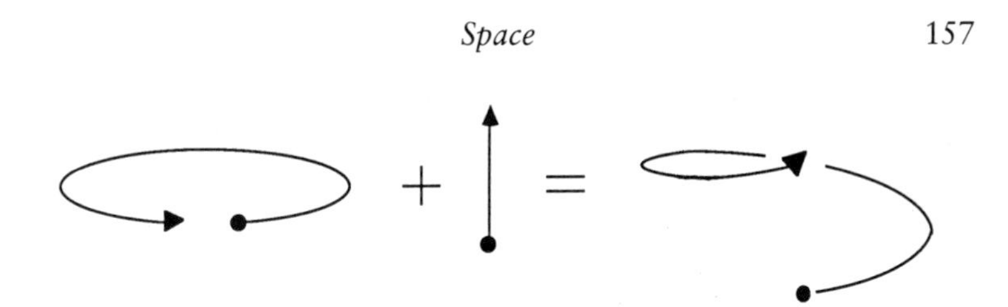

Fig. 85 Circle + line = helix.

more closely, I see that the bump has a texture something like the aerial view of a weathered mountain range. Closely regarding one of the bump's individual peaks, I note that the little peak's slopes are wrinkled from the drying of the paint. If I get out a magnifying glass, I find small cracks in the paint; a microscope reveals saw-toothed edges along the cracks; an electron microscope would break the saw teeth into paint molecules, and so on. And this is the wall that I would have liked to think of as part of a plane!

Each part of the real world around us contains an inconceivably large amount of information. For all practical purposes the world is, in each of its parts, infinite. Anything you look at manifests more and more structure as you look closer. Abstractly speaking, the form of any object exists independently of the object. This is the software–hardware distinction. The actual particles making up that bump on my wall serve as the hardware — the material that codes the pattern up. The idealized pattern that is being coded up is the bump's software. Here I am using software to mean something like form, and hardware to mean something like content. If I know the bump's software well enough, I can construct an exact replica of it whenever I want to. Given that an object exists, its form exists, so it is legitimate to reason about form as a definite concept. My point is that talking about endlessly complex forms is a legitimate thing to do.

Now, mathematics is the science of forms. Although mathematics begins with very simple forms, very complex forms do exist, and mathematics should be able to say something about them. The problem is in finding something interesting to say!

Of course, one way to increase the complexity of the space forms we look at is to include more variables; that is, to think in terms of higher dimensions. Just as algebraic curves are given by polynomials in x and y, algebraic surfaces can be described by polynomials in x,

y, and z. There are well-known surfaces that correspond to most of the sorts of curves we've already discussed. The simplest surfaces are the quadric surfaces, which correspond to xyz equations of the second degree.

A different approach to increasing complexity is to stick to curves and look at curves of infinite degree. The sine curve and the exponential curve can be represented as endless power series that are, in effect, polynomials of infinite degree. Curiously enough, such power-

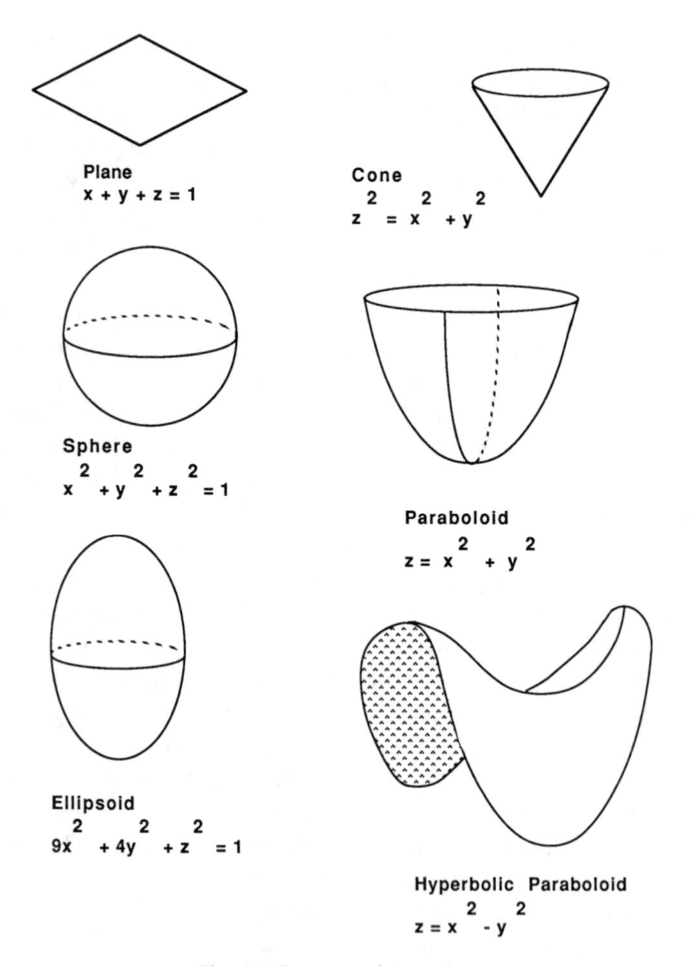

Fig. 86 Some quadric surfaces.

series curves are not anywhere near as complex as such real-world curves as the contours of a fish, the profile of a mountain range, etc. Power-series curves don't have any corners, and tend to change in smooth and unsurprising ways. An important theorem, discovered by Brook Taylor in 1712, says that if you know exactly what a power-series curve looks like at any one point, then you can predict exactly what it looks like everywhere else. The curve is given, if you will, by a single "infinite bit."

A more effective way of getting infinitely complicated curves is to allow our curves to have corners in them — to piece curves together from different glued-together pieces. An extreme form of this kind of process occurs in Peano's space-filling curve of 1890 (Fig. 88). The Peano curve, by being "all corners," is so jagged that it fills a whole

Fig. 87 Brook Taylor sees enough.

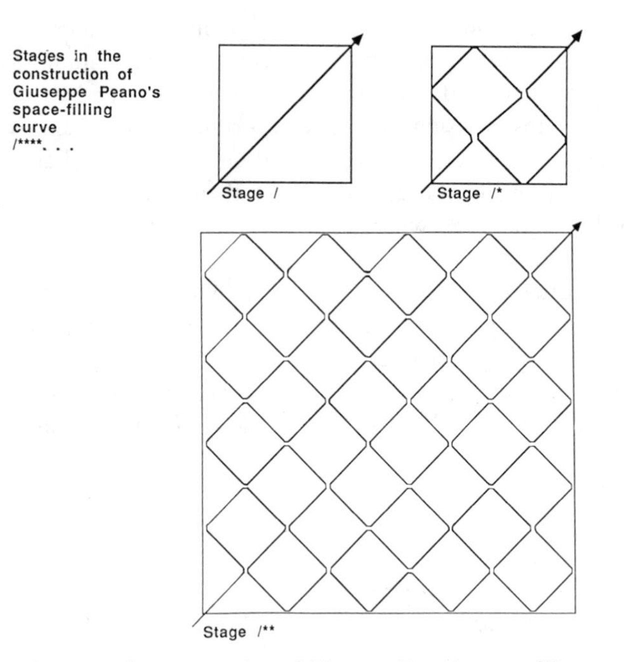

Fig. 88 Stages in the construction of Giuseppe Peano's space-filling curve:
/****_ _ _   /****_ _ _   /****_ _ _

square with a dense, two-dimensional scribble. The curve is constructed as the limit of endlessly many stages, which I call /, /*, /**, /***, /****, and so on. The starting stage, /, is simply a line segment cutting diagonally across the square. Each of the successive stages is gotten by repeating the operation "*," which consists of replacing each straight segment by a kind of zig-zag or S that goes LRRRLLLR, where "L" is left and "R" is right.

Most mathematicians initially viewed Peano's space-filling curve as a monster, but more and more shapes like this began appearing. Topology, which is the study of shapes, came up with a variety of endlessly complex "monsters" in the 1930s and 40s. Perhaps the most picturesque topological monster is Alexander's horned sphere, discovered in 1924 (Fig. 89).

In the 50s and 60s, Benoit Mandelbrot developed a new way of looking at certain infinitely complex shapes. He named them "fractals."

Fractals

Which is simpler, a circle or a polygon? It depends on how you look at it. From a number-oriented viewpoint, a polygon is simple and a circle is complex. From the viewpoint of number, a polygon is gotten by the easy method of hooking together a certain number of equal line segments, whereas a circle arises only as an ideal limit — a polygon with infinitely many infinitesimal sides.

From a space-oriented viewpoint, a circle is simpler than a polygon. From the viewpoint of space, a circle is gotten simply by taking a radius r and a central point C, and letting the circle be all the points P that lie at a distance r from C. In space a polygon can only be given by finding some way to separately specify each of its corners.

If we use algebraic equations to talk about space curves, we find that a circle has a very simple formula, while a polygon has a much more complex kind of formula, with many terms, one term to define each of the polygon's sides.

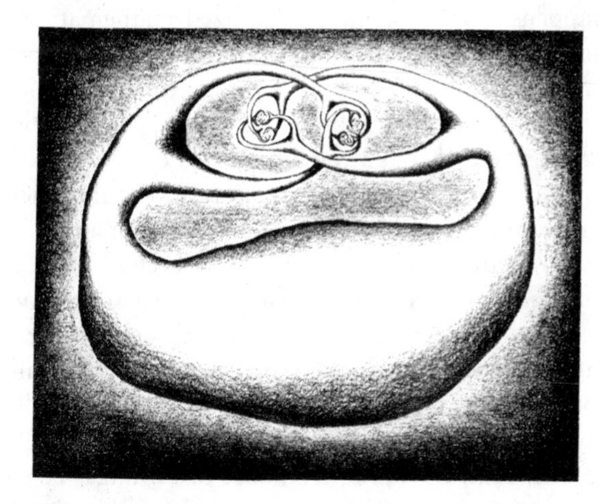

Fig. 89 Alexander's horned sphere. From John Hocking and Gail Young, *Topology*, © 1961, Addison-Wesley, Reading, Mass. Page 176, Fig. 4–11. Reprinted with permission.

In this section I will begin by talking about curves made of branching line segments. In some ways these branching curves are like polygons. Then I will go on to talk about the class of curves that arises if we look at branching curves that have infinitely many forks and arbitrarily small branches. There is a sense in which the endlessly branching curves are simpler.

If you look at trees during the winter, you will notice a number of things about their methods of branching. One of the most interesting facts is that most trees are roughly *self-similar*. This means that at each level, the number of branches coming out of a fork is the same. Some trees, like elms, have two branches coming out of most forks; some, like oaks, have several branches coming out of most forks.

The point is that in any given tree one of its branches can serve as a good scale model of the whole thing. The large, overall branching pattern of a tree is repeated along each of its branches. For this reason, trees, like snail shells, can be thought of as self-similar. The same overall pattern is repeated at various different size scales.

Of course, once you get down to a tree's smallest twig, you no longer have any branchings at all, and the self-similarity is gone. But before the twig level is reached, most big trees will have gone through some seven forks, or seven levels of self-similarity.

What happens if we imagine an idealized mathematical tree that branches infinitely often? The simplest tree like this is known as "the infinite binary tree." If each branch is taken to be half as long as the preceding branch, the whole infinite binary tree can be fit into a unit square. Any path out through the binary tree can be described as an endless sequence of zeros and ones, and it turns out that each of these paths leads to a unique point on the unit line segment at the top of the square. Since there are endlessly many levels beyond any given fork, cutting the tree at any fork yields an exact scale model of the whole tree. The infinite binary tree is perfectly self-similar.

So far as we know, none of our real-world shapes really has an infinite amount of detail. But just as a circle is conceptually simpler than a polygon, an infinitely branching tree is conceptually simpler than a tree with some fixed, finite number of forkings. To specify the infinite binary tree, we simply say, "start with a pair of branches and repeat step (+) forever," step (+) being "Replace each branch's endpoint with a pair of branches half as long."

Carrying out this type of process in various ways leads to a variety

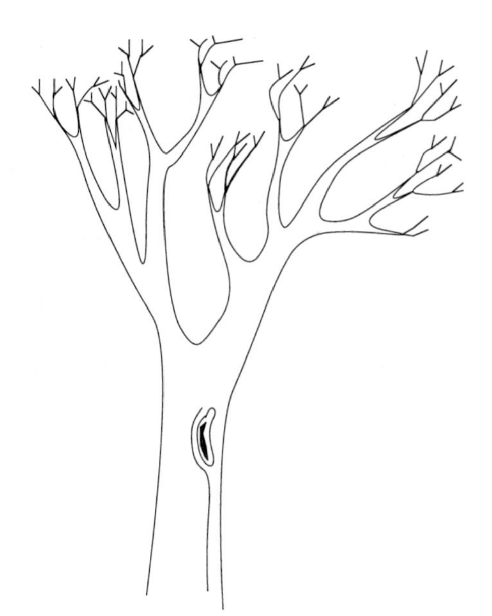

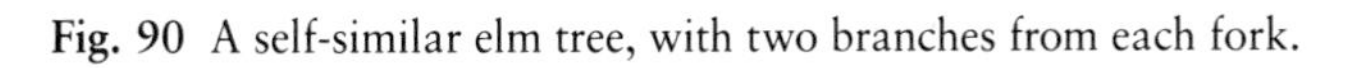
Fig. 90 A self-similar elm tree, with two branches from each fork.

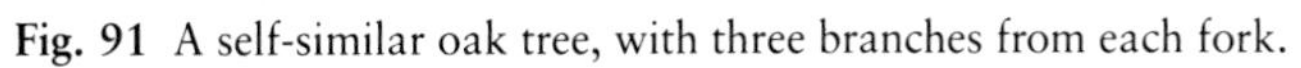
Fig. 91 A self-similar oak tree, with three branches from each fork.

of infinitely detailed, perfectly self-similar shapes called "fractals."

The modern theory of fractal shapes is primarily the creation of one man: Benoit Mandelbrot, a French-born mathematician now based at IBM's research center in Yorktown Heights, New York. Mandelbrot's basic theme is stated on the first page of his 1977 book *The Fractal Geometry of Nature:* "Clouds are not spheres, mountains are not cones, coastlines are not circles, and bark is not smooth, nor does lightning travel in a straight line. More generally, I claim that many patterns of Nature are so irregular and fragmented, that, compared with . . . standard geometry, Nature exhibits not simply a higher degree but an altogether different level of complexity."

Mandelbrot was struck by the fact that, depending on the scale of map one uses, a country's coastline looks quite different and has quite

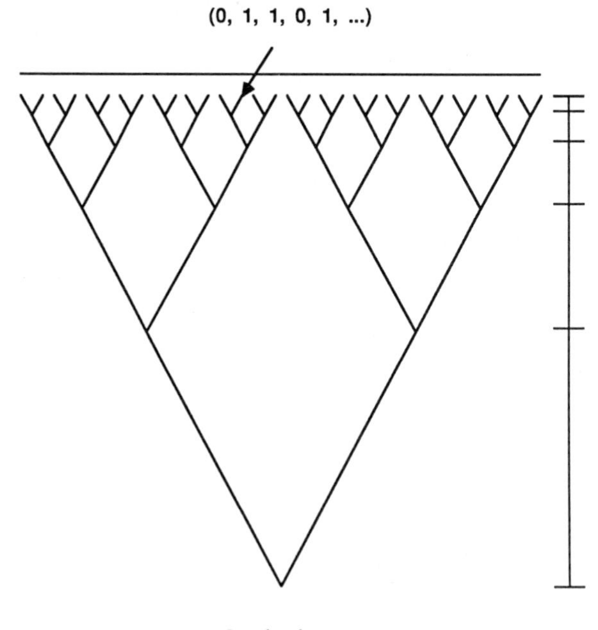

Fig. 92 The infinite binary tree.

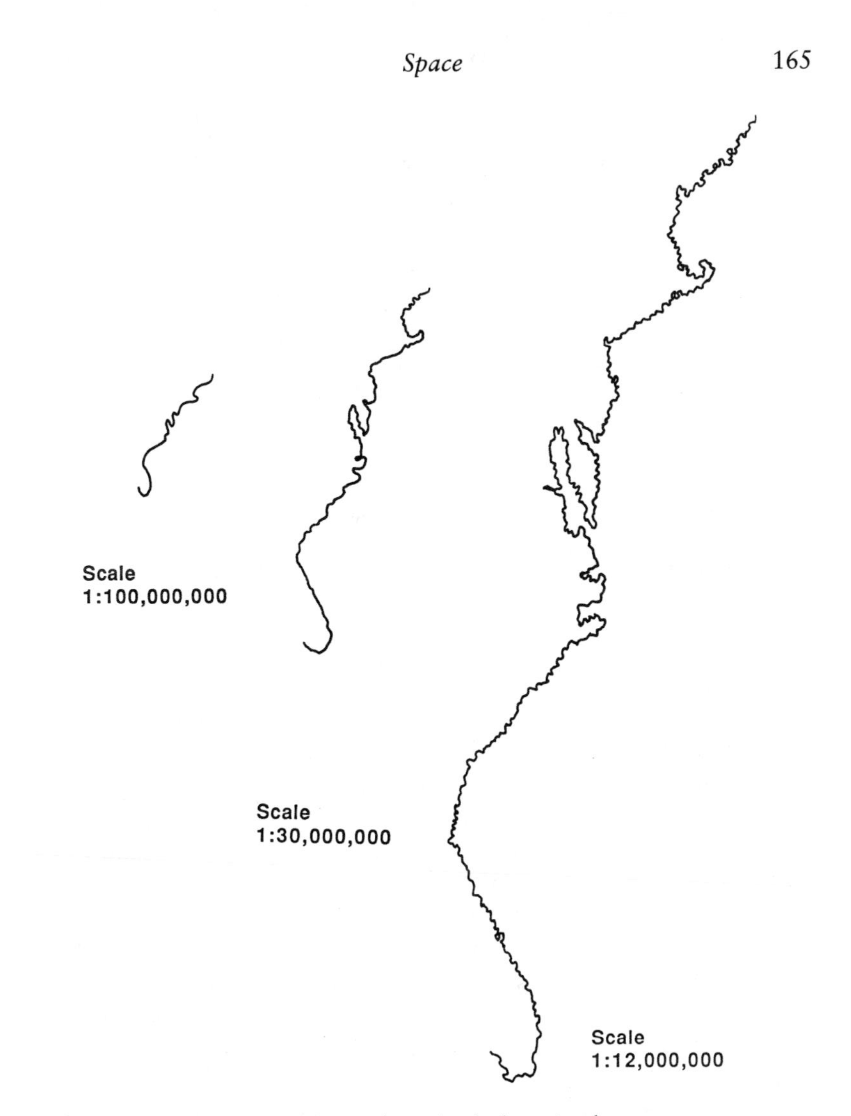

Fig. 93 The larger the scale, the bumpier the coast.

different lengths. On a world globe, the eastern US coast looks like a fairly smooth line some two or three thousand miles long. On a large-scale map of the US, this same coast looks much more jagged, and adding in the lengths of all the capes and inlets, the total length

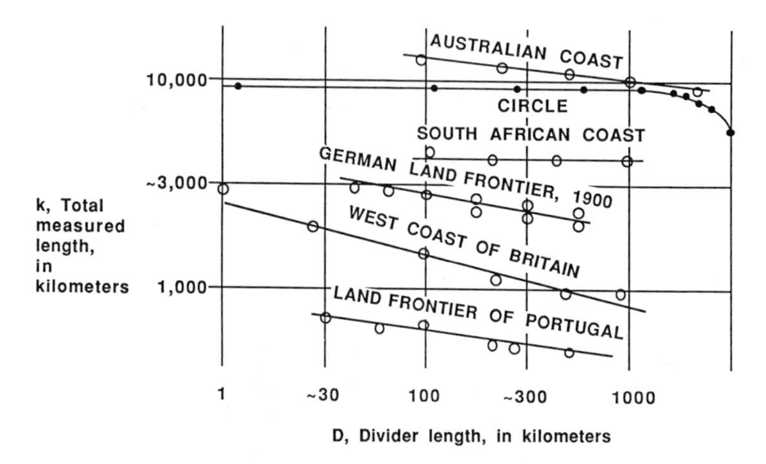

Fig. 94 Mandelbrot's plot of Richardson's data. Adapted from Benoit Mandelbrot, *The Fractal Geometry of Nature,* W.H. Freeman, San Francisco, 1982.

seems more like four or five thousand miles. If one pieces together a giant coastal map from detailed navigational charts, the coast becomes an incredibly complex curve ten or twelve thousand miles long. If you set out on foot to walk the coastline yourself, staying always within one step of the water, the many little zigzags too small for a map will build the length up to perhaps fifteen thousand miles. If a determined ant were to walk the coastline, staying always, let us say, one ant length from the water, the countless detours around rocks and pebbles might push the coastline's length up to thirty thousand miles, and still smaller coastline explorers would find the total length ever longer.

Mandelbrot's initial studies of fractal curves were inspired by some studies that the British scientist Lewis Fry Richardson did on the way in which a natural border's length depends on the value of the unit one uses to measure it. Mandelbrot represents Richardson's results on a graph in which the value *D* of the measuring unit and the value *k* of the total length gotten are represented as length on number-log lines (Fig. 94). Note that along a smooth curve, such as a circle of longitude, the value of *k* soon stabilizes at a fixed value, but for bumpy coastlines, the value of *k* seems to go up indefinitely as the value of *D* shrinks.

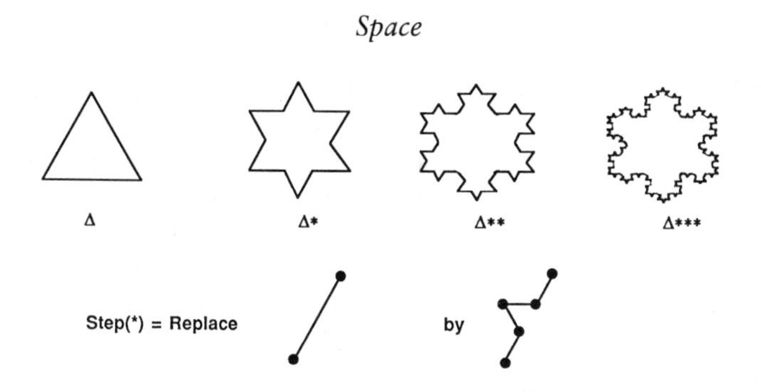

Fig. 95 The first four stages of the Koch curve, $\triangle^{****}\cdots$.

The point of this train of thought is that such seemingly simple notions as "the length of a curve" seem not to apply to naturally occurring curves like the US coastline. Yet the coastline really exists — it is the boundary between land and water. The lesson is that there is something wrong with trying to model our coastline with anything like an ordinary algebraic curve. A new mind tool is called for.

Mandelbrot's *fractal curves* are lines that have zigzags at every size scale. The tree patterns we looked at above are very simple kinds of fractal curves, because the different-sized branches occur in different spots. In a general fractal, zig-zags of every size occur everywhere on the curve.

Defining such a general fractal curve is not so difficult as one might think. In 1906 the mathematician Helge von Koch showed how to construct a continuous "snowflake" curve, with infinite perimeter, that bounds a finite area. The idea is to start with an equilateral triangle, and then to endlessly repeat a step that we might call "(*)": Replace each straight line segment with a segment with an equilateral-triangle-shaped bump on its middle third."

The idealized mathematical limit of this process is a perfectly self-similar curve of infinite length, a curve that bounds a complex shape of area eight-fifths the area of the original triangle. The curve itself becomes so fuzzy that it is no longer meaningful to speak of it as a one-dimensional line. According to Mandelbrot, the Koch curve should be thought of as having a dimension of 1.2618 ($= \log 4/\log 3$). It is this *fractional* dimension value, as well as the curve's *fractured* appearance, that led Mandelbrot to adopt the term "*fractal* shape."

But how, one wonders, can a curve have a dimension other than one? A crude way of thinking of it is as follows. When I say an ordinary curve has dimension one, I mean that, so long as I am on the curve, I have only one degree of freedom in my motion: I can speed up or I can slow down. This is as opposed to being on a curved surface, such as a hillside, where I have two degrees of freedom: I can speed up or slow down, veer to the left or to the right. Now, if I am walking along an approximately fractal curve such as a coastline, I can speed up or slow down, as before, but now I also have the freedom to veer slightly to the left or to the right, depending on how closely I can stick to the water's edge. I am constrained by the water's actual curve, but I am freed by my ability to look at different size scales. If the curve is spiky, the fractional dimension is high; if it is smooth, the dimension is low.

It is worthwhile to investigate Mandelbrot's measure of fractal dimension in a bit more detail. Suppose we go back to the notion of measuring a curve by placing a pair of dividers down on it over and over. Let the separation of the dividers be D. Let the number of times that the dividers fit onto the curve be N. The value for the curve length arrived at by this process is $k = N \times D$. If the curve happens to be a straight line k units long, then this value is in fact correct. We can rewrite our equation as $N = k/D$.

The length k is simply the number of times that a one-unit divider fits along the line. Making the divider, say, three times as small leads to three times as many divisions. If we write N as $N(D)$ to express N's dependence on D, we see that $N(1) = k$, $N(1/3) = 3k$, $N(1/9) = 9k$, and so on.

In the process of studying his graphs of Richardson's data, Mandelbrot got the notion that on a fractal curve, a different equation holds, $N = k/(D^a)$. Again, k is the number of times that a unit-length divider fits along the curve, but if the curve is fractal, making the divider three times as small may lead to *more than three* times as many divisions. Specifically, recall that, on the Koch curve, making the divider three times as small gives *four* times as many divisions, and making it nine times as small gives sixteen times as many divisions.

Qualitatively, we can see that this effect will be accomplished if a has a value greater than one. Here $N(1) = k$, $N(1/3) = (3^a)k$, $N(1/9) = (9^a)k$, and so on. Mandelbrot calls a the "fractal dimension."

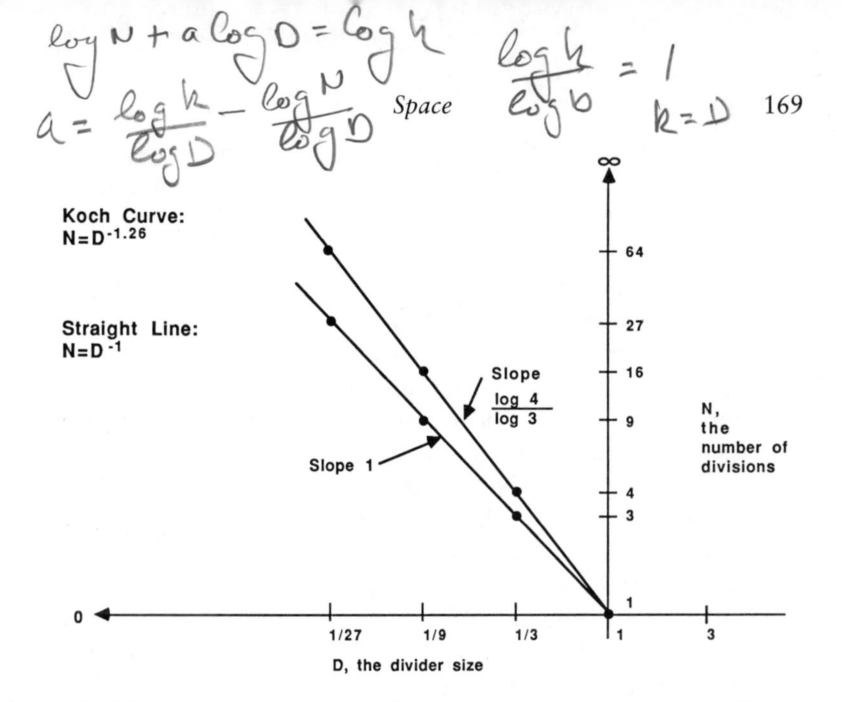

Fig. 96 A number-log versus number-log graph comparing detail in a line and a fractal (numbers represented by sizes of their log3).

Since on the Koch curve we want to have $N(^1/_3) = 4k$ and $N(^1/_9) = 16k$, we want the fractal dimension a of the Koch curve to have the property that $(3^a) = 4$. A little logarithmic sleight of hand shows that this means that a should be $\log 4/\log 3$:

$$\begin{aligned} 3^a &= 4, \text{ so} \\ \log 3^a &= \log 4, \text{ so} \\ a \log 3 &= \log 4, \text{ so} \\ a &= \log 4 \,/\, \log 3. \end{aligned}$$

In general, if shrinking the measuring unit by a factor of P increases the number of units along the curve by a factor of Q, then the curve has fractal dimension of $\log Q/\log P$.

Usually we imagine that if we pay twice as much attention to something, we'll get twice as much information, but imagine a fractal body of information. Imagine a situation where paying twice as much attention gives you *three* times as much information. In general, a fractal with dimension $\log Q/\log P$ has the property that examining it P times as closely gives you Q times as much information, with Q greater than P! The old truism "God is in the details" is true with a vengeance for fractals.

The same definition of fractal dimension carries over to surfaces. Instead of measuring a surface with little measuring sticks, one usually measures it with little squares. To cope with the problem that a measuring stick touches a fractal curve at a variety of isolated points, we thought in terms of dividers. By the same token, we might think of measuring surfaces in terms of things with four prongs arranged in a square, something like the business end of a corn-cob holder.

A measuring square with edge D^2 has an area of D. If we have some flat square with area k, then the number N of measuring squares that fit onto the flat square will clearly be given by the formula $N = k/D^2$. Now k is the number of times that a unit square fits onto the flat square. In other words, $N(1) = k$. Suppose we take smaller dividing squares, $N(^1/_2) = 4k$, $N(^1/_3) = 9k$, and so on. The number of divisions varies as the reciprocal of the second power of the square's edge.

A naturally occurring surface like a mountain is irregular at many

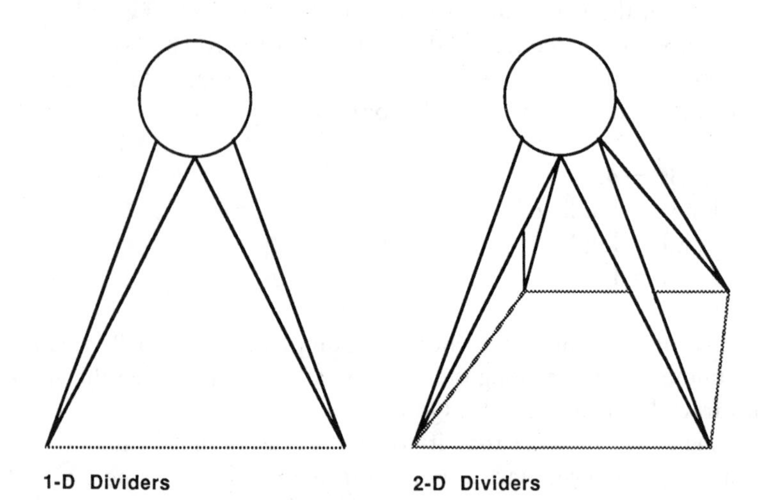

Fig. 97 Approximating length and area.

different size scales, and we would expect the number of measuring squares on such a surface to increase faster than it would on a flat surface. We expect that there is a value a greater than two such that $N = k/(D^a)$. As before, if shrinking the edge of the dividing square by a factor of P leads to Q times as many divisions, we take the fractal dimension a to be $\log Q/\log P$. In a flat square, Q is simply (P^2), and as expected,

$$a = \log Q/\log P = \log(P^2)/\log P = 2 \log P/\log P = 2.$$

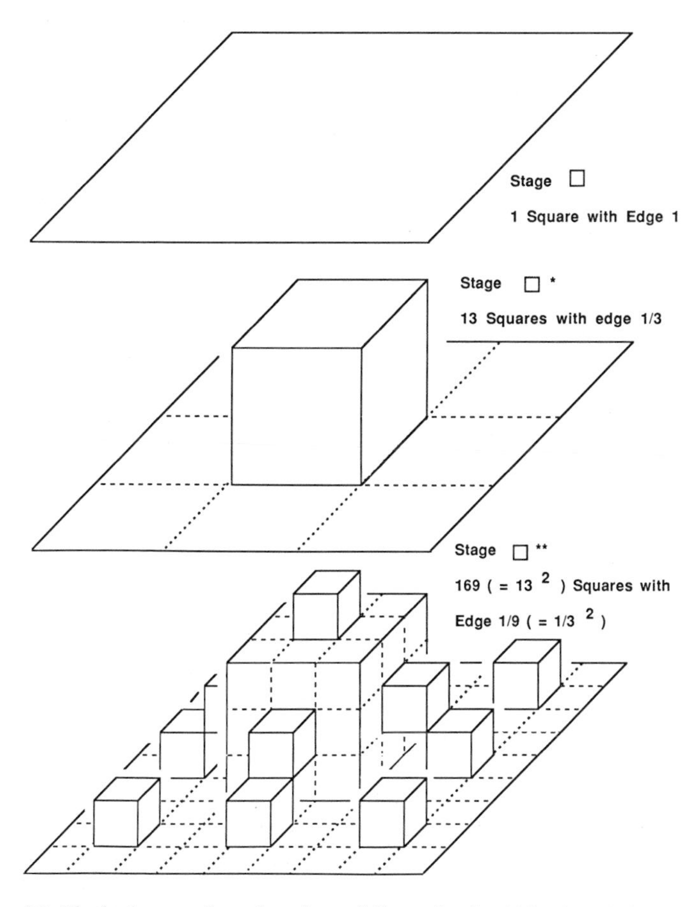

Fig. 98 The hydrant, a fractal surface of dimension log13/log3 = 2.334.

One way to construct a regular fractal surface is to keep adding cubes to the middle square. This leads to a fractal surface with dimension log13/log3, which is 2.33. The surface might be called a "hydrant." If the edge of the starting square is taken as a unit, then in the limit, the hydrant's center is exactly $^1/_2$ unit high, the eight little hydrants around the edge are each $^1/_4$ unit high, and so on. If we construct a hydrant on each of the six inner surfaces of a unit cube, they just fit, and we get a 2.33-dimensional surface that might be called the "hydrant cube."

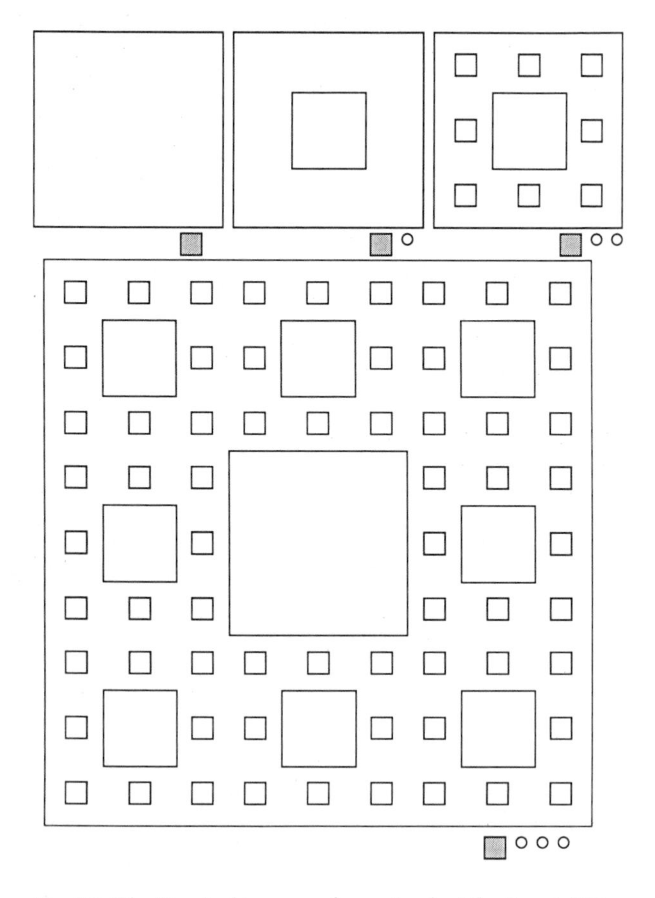

Fig. 99 The Sierpinski carpet, dimension log8/log3 = 1.893.

If we only look at the part of the hydrant surface that lies on the bottom plane, we get a fractal known as a Sierpinski carpet. The carpet can be defined on its own as the limit of a process of repeatedly removing the central $^1/_9$ of squares. The carpet's fractal dimension is log 8/log 3, about 1.9, so we might better think of it as a very bumpy curve than as an actual surface. Since the carpet is, again, the intersection of a plane with the hydrant surface, it is indeed reasonable to think of the carpet as a curve.

The Menger sponge is a higher-dimensional analogue of the Sierpinski carpet. It is created by starting with a solid cube and repeating this step: Divide into twenty-sevenths, and remove the central cube as well as the cubes lying at the center of each face. Seven cubes are removed in all, so looking three times as close shows twenty cubes remaining, which makes for a fractal dimension of log 20/log 3, or 2.727. The Menger sponge is best thought of as a complicated surface. As it hap-

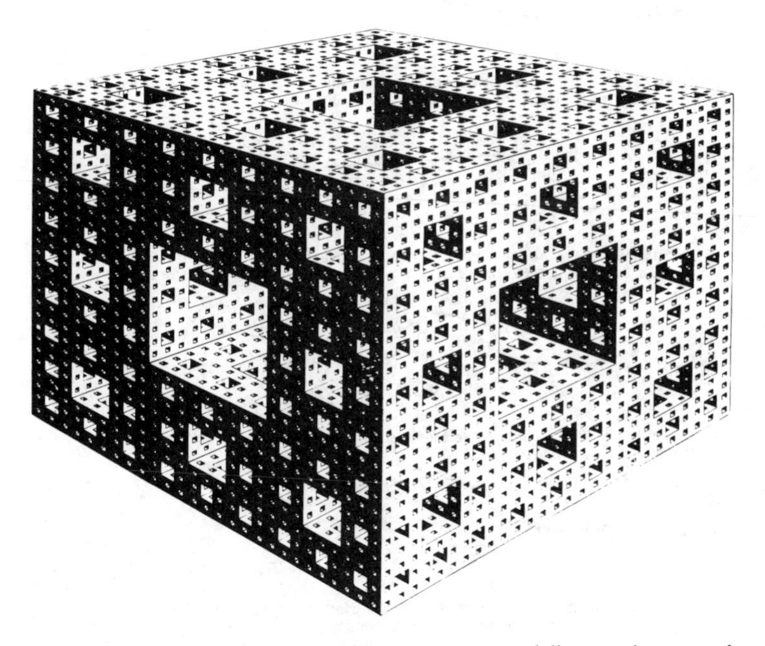

Fig. 100 The Menger sponge. From Benoit Mandelbrot, *The Fractal Geometry of Nature*, W. H. Freeman, San Francisco, 1982.

pens, the picture of a Menger sponge in Fig. 100 could equally well be a picture of the hydrant cube. The difference is that in the Menger cube the central cubes (which are not visible in this projection) are all missing. The hydrant cube actually fits inside the empty spaces of the Menger sponge. One might think of the sponge as being a vast department store and the hydrant cube as being an influx of shoppers!

A difference between a Koch curve and an actual coastline is that the Koch curve is an idealized mathematical form with structure at *infinitely* many levels. Just as we do not think of a real tree as having

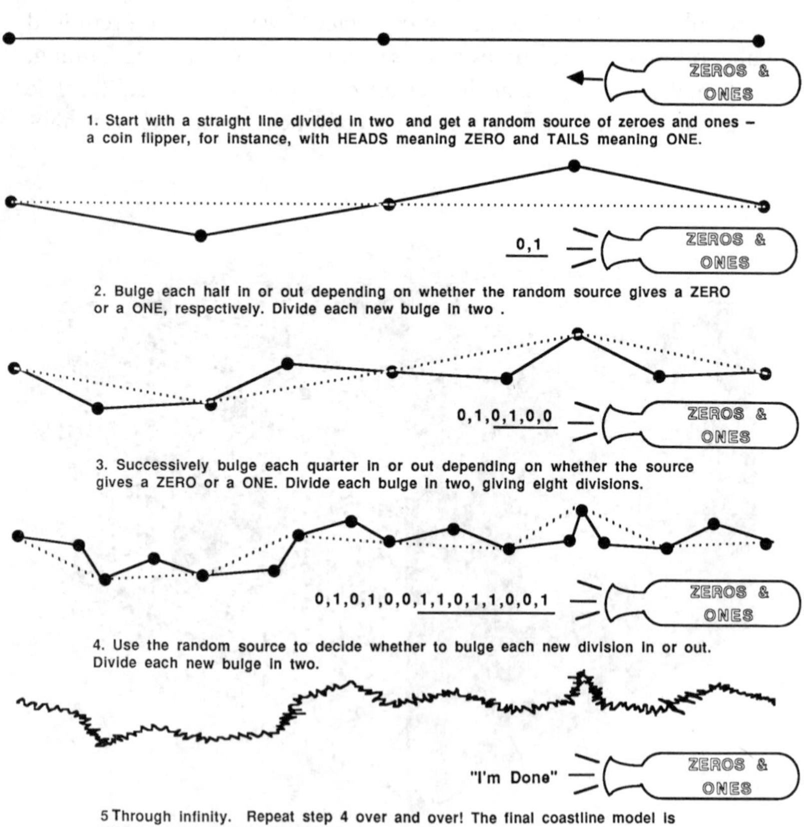

Fig. 101 A random fractal.

infinitely many forks, we do not think of a physical coastline as really being an infinitely complex line, but for many purposes, a fractal is a better model than is a finitely complex smooth curve. The drawback in using a fractal curve as a coastline model is that the fractal has too much detail; the weakness in using a finitely complex curve for a coastline model is that the finitely complex curve is so much harder to describe.

Another objection to using a curve such as the Koch curve as a coastline model is that the Koch curve is so regular. By magnifying, or scaling up, we can convert a small piece of the Koch curve into a replica of one of its larger pieces. It is perfectly self-similar. We do not expect our real coastline to be so regular. It turns out that we can construct a kind of irregular Koch curve by randomly alternating between pointing the bulges out and pointing them in.

This kind of random fractal curve construction corresponds rather well to the way we imagine the coastline to have actually been built up. On the one hand there are forces acting at many different size scales; this gives the curve its fuzzy, fractal nature. On the other hand, the forces are acting randomly; this gives the curve its overall meandering appearance.

If we randomize the construction of a branching-tree fractal, some very realistic-looking forms arise. If kept flat, such tree patterns may bear a resemblance to the overall pattern of a drainage basin — a river, its tributaries, the creeks that run into the tributaries, the streams that run into the creeks, etc. Other natural objects with forms like fractal trees are the lung's bronchial tubes and the circulatory system's network of veins and capillaries.

The process that leads to a generalized random fractal curve can be used to produce generalized fractal surfaces. One divides a given area into successively smaller regions and in a random way bulges these little regions up or down. One of the most visible achievements of Mandelbrot's researches into fractal forms has been his ability to create computer graphics that look like realistic landscapes. These forms are beginning to be used in science fiction movies as alien planetscapes. Mandelbrot's concept of fractals as fundamental to physical reality is supported by the fact that his patterns really *do* look natural.

We have talked about fractal curves and fractal surfaces. A special

Fig. 102. Computer-generated fractal surface. From Benoit Mandelbrot, *The Fractal Geometry of Nature,* W. H. Freeman, San Francisco, 1982. Actual implementation by Richard F. Voss.

type of fractal solid can be gotten by starting with a solid block of matter, dividing it into a stack of smaller blocks, randomly removing the matter from some of the smaller blocks, subdividing the smaller blocks into still smaller blocks, and repeating the process. This leads to a spatial distribution of matter known as a "fractal dust." The water droplets in a cloud resemble a fractal dust, as do the stars and galaxies in outer space.

It is also possible to imagine situations in which fractals of different sorts are layered together. Consider, for instance, an elm tree. As a physical object, it is more than a simple fractal regress of lines. For about seven levels it's based on lines, then the size of the surface cuts in. Bark is a fractal based on 2-D units. Viewed closely, a piece of bark is like a plain with a canyon in it — the walls of the canyon are smooth, but with cracks in them, and the cracks have crevices in their walls, and so on for perhaps seven further levels. At this point we get to cell size, and the fractal becomes based on 3-D units. Every large-scale micrograph shows that cells are filled with fractal structures: dusts, trees and surfaces. Presumably going further and further down will lead to yet more fractal structures, with all kinds of dimensionalities.

For thousands of years, mathematicians have been unable to really grapple with the kinds of shapes that the world actually contains. We have had, in effect, no precise words for shapes like seaweed. The Greeks spoke of endlessly complex shapes as "apeiron," meaning formless. This can be taken literally, for if we have a really clear image of a form, then we can express this form in mathematical terms. If I can't organize my ideas about some physically occurring pattern of information into a precise mathematical framework, then I cannot be said to have a really clear understanding of the information in question, and in this sense the pattern has no form. A mathematical description of a real-world object can be thought of as a short code for a complicated pattern.

Speaking casually, I say that a shape is "fractal" if it has similar-looking structures on several different size scales — a line that branches into lines that branch into lines that branch; a bump covered with bumps that are covered with bumps that are bumpy; a glob made of globs made of globs of globs. The more levels there are, and the more the levels look like each other, the more perfect a fractal I have. Another characteristic feature of fractals is that they have *more* detail than expected. Typically, examining a fractal twice as closely will give you more than twice as much information. In the next section I will argue that my knowledge of the world falls into a pattern that is fractal in nature.

Life Is a Fractal in Hilbert Space

Suppose that I want to write down a complete description of some part of my life. I am going to argue that such a description must ultimately take the form of a fractal in a multidimensional space.

Let me take a simple example. While thinking about fractals, I pick up my lit cigarette and inhale, exhale. We're talking about a thirty-second time period, from 4:08:30 thru 4:09:00. Simply to say, "I picked up a cigarette and took a puff" in no way provides an exhaustive description of what those thirty seconds are like. Being alive is not really a linear process. Being alive, and thinking, is multidimensional. How to write it down?

The notion of a menu-driven computer program can come to our aid. I'm thinking about a new type of text. Suppose that I were to create a special software that had the following features:

1. At any given time a screen of text is displayed. The "reader" can move out of the given screen in a number of ways. Just to have something concrete to think about, let us suppose that the starting screen looks like this:

> It was Tuesday, July 3, 4:08. I was sitting in front of my computer screen, working on a book. I'd just written about fractals, and now I had the idea of trying to create a fractal text. I picked up a lit cigarette from the ashtray to the right of the keyboard, and took a puff.
>
> PUSH "NEXT SCREEN" FOR MORE, OR MOVE CURSOR TO A WORD AND PUSH "MENU."

2. One way of moving out of a given screen is to call for the next screen. This corresponds to moving forward along the natural line of thought, time, or association in which the text was written.

3. Another way of moving out of a given screen is to move the cursor to any of the words displayed and push the **MENU** button. (The cursor, for those not familiar with computers, is a bright dot that can be moved around the screen by pushing some keys with arrows on them.)

4. When you push the **MENU** button for a particular word, the screen displays a number of options. Each of these options corresponds to a kind of dimension of knowledge. Thus, if you move the cursor to "cigarette" and push **MENU**, you might get the following screen:

( ) The TYPE of cigarette I smoke.
( ) My FIRST smoking experiences.
( ) Facts about TOBACCO.
( ) My feelings about CANCER.
( ) OTHER SMOKERS I know.

SELECT AN OPTION BY POSITIONING CURSOR AND PUSHING "RETRIEVE," OR PUSH "NEXT PAGE" FOR MORE DETAILED OPTIONS.

5. Selecting an option on a menu screen gives a new screen. Say, for instance, that you were to move the cursor to "My FIRST smoking experiences." Pushing **RETRIEVE** would give the following screen:

One of the first times I remember smoking was at a baseball game in Louisville. My friend Mike Dorris and I had a pack of Winstons, and we seated ourselves in an isolated part of the bleachers so that no grown-ups would yell at us. We were about fourteen. Neither of us was interested in baseball. Mike showed me how to "French inhale," meaning that you let the smoke trickle out of your mouth and suck it in through your nose. The cigarette smoke made us dizzy and high. . . .

PUSH "NEXT SCREEN" FOR MORE, OR RETURN TO PREVIOUS SCREEN BY PUSHING "UNDO," OR MOVE CURSOR TO A WORD AND PUSH "MENU."

6. In general, you can move out of any given text screen in three ways: (a) you can push **NEXT SCREEN** for more material along the line of thought presently under way, (b) you can move the cursor to a word that interests you and then push **MENU**, or (c) you can push **UNDO** to go back "up" the branch to one of the screens you started from.

The text described by this process would have a structure something like a fractal tree. What might I call it? "My life story" would not be quite accurate, since it would not have the linear quality that one expects from a story. I think it might be better to call it "my life fractal."

We can think of the original narrative line as a string of words.

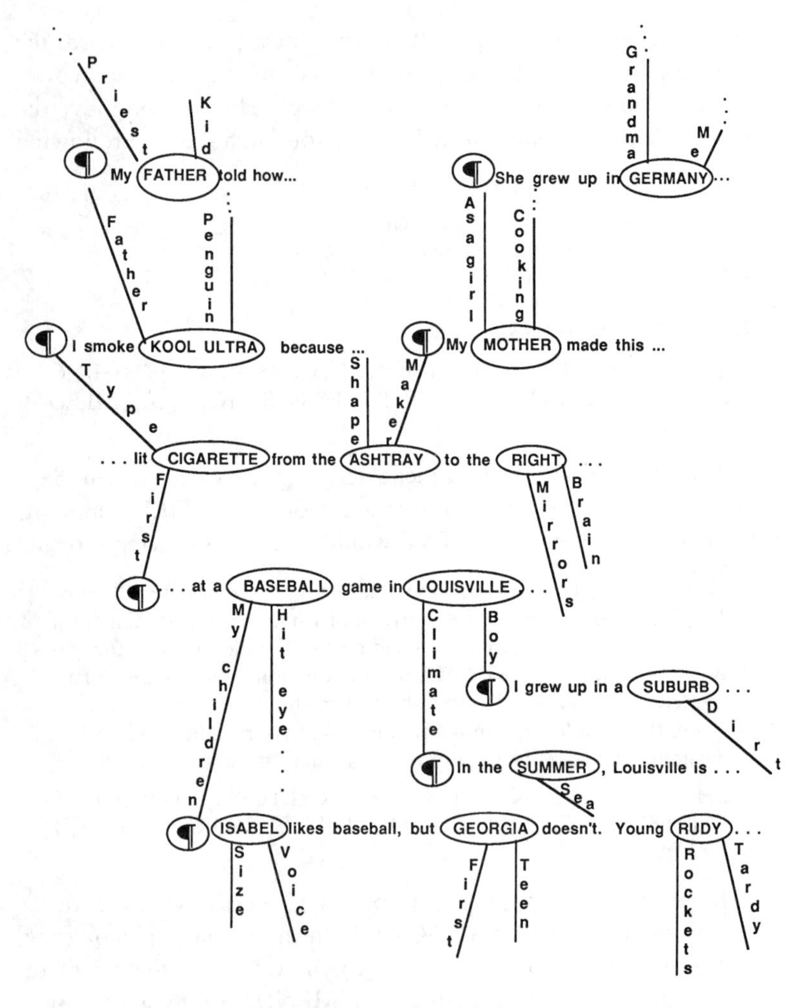

Fig. 103 Part of my life fractal.

Simply by pushing **NEXT SCREEN** whenever necessary, one can proceed along a given string as long as one likes. There are branches going off from each important noun, verb, or unified phrase. One uses the **MENU** screens to select which branch.

Each branch, in turn, leads to a new string of words, and there are further sub-branches going off from each of the key words in the new string. Ideally, the branching continues without end.

If at any time you want to trace back to a branch point you went through, you push the **UNDO** key until you get back. Rather than being a straight line, a path through such a narrative might look something like the path shown in Fig. 104.

Recall that a characteristic feature of a fractal is that looking closer reveals an unexpectedly large amount of detail. A software describing my life would ideally have this quality, in the sense that the closer one examines a concept, the more branchings from it one can find.

Accessing my life fractal in your computer would be very much like talking to me over a teletype line. It would be more or less the same as having a conversation with me, for how does a conversation operate, if we happen to be in a situation where you are interested in finding out about me (as opposed to telling me about yourself)? I start talking about something, and after a while you interrupt and say something like, "What were your early experiences with smoking cigarettes?" I start talking about that, then maybe after a while you say, "You mentioned baseball a minute ago. Do your children like baseball?" And so on.

Someday I hope I have enough time to really get a good piece of my life fractal down on disk. Of course, my life fractal already exists as a software pattern coded up in my brain, but if I could make a bunch of copies of it, it would be a step toward software immortality, and my work would be that much easier to build on, in case anyone wanted to.

Obviously I'm not alone in this wish to record my life fractal — my personality. I would hazard a guess that, in the future, recording one's own life fractal will be a very popular activity among retired people. Already it is not uncommon for an older person to write down

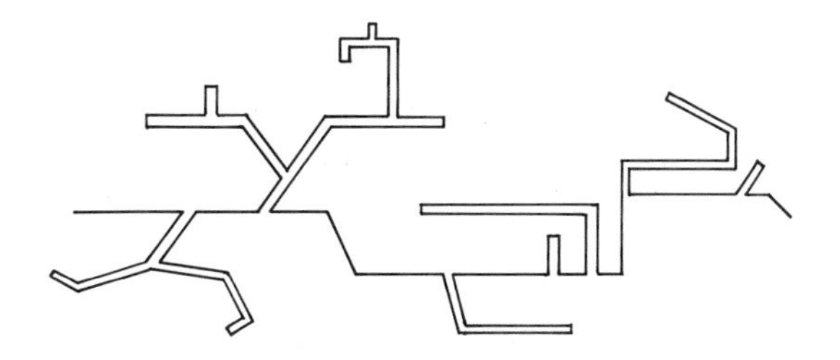

Fig. 104 A trajectory through my life fractal.

the "story of my life," but how much easier and how much more complete it would be if one could purchase a small computer, called, let us say, a life box. You could tell random reminiscences to your life box and it would store them all and set up a system of cross-referencing. Occasionally the life box might ask you a question to clear up certain confusing links. After a few months it would have you down in some detail, and you could pass it on to your children so that your grandchildren would be able to hear for themselves "what Grampa was like." Who wouldn't want to immortalize himself this way? The first person to design a really good life box will become a millionaire.

Why exactly do I describe this kind of software construction as a fractal? The reason is that any one word in the pattern is really about as central as any other word. At each position there are a whole lot of branches leading off, and this really is the way it feels, isn't it? If someone wants to talk to me about myself, I can start just about anyplace. Indeed, rather than always having the program start up with the 4:08 cigarette lighting, it would be just as good to have the program start at a random position, or with the branches going off of any particular word that the reader cares to select.

How many branches will go off from each key word? Very many. If one takes more and more refined qualities into consideration, there is no end to the questions one can ask about a concept as simple as "CIGARETTE." Any given concept in my world relates to many, many other concepts. Pushing it a bit, we might even say that there are infinitely many branches at each point. As I will explain in the next section, mathematicians use the phrase "Hilbert space" to stand for a commonly used type of infinite-dimensional space, so if we think of each line of inquiry (branch direction) as being something like a dimension, it makes sense to speak of my life fractal as a fractal in Hilbert space.

Viewed from the outside, I am really the same thing as my complete life fractal. Inside my life, I have feelings of time passing, of being alive, of having free will, and so on, but in terms of what I can actually tell about my life, I am the same as my software, the same as my life fractal.

Hilbert space, with all its dimensions, has room for lots of things. It has room for your life fractal as well as mine; it has room for *everyone's* life fractal. It seems evident that these fractals will connect in certain spots, for instance at spots where people interact, or at spots where people share common beliefs. It all adds up to one big universal life fractal!

What makes this an interesting way of looking at life is that it stresses that life is, above all, a mental phenomenon. The hardware is really almost incidental. What counts is the wondrously intricate tangle of the big life fractal that our softwares make together.

Hilbert Space

Recall that something is said to be digital if it comes in discrete, separate lumps; while something is analog if it is smooth and continuous. Singing is analog; Morse code is digital. Rheostat dimmers are analog; on–off switches are digital. Watches with dials are analog; watches with little number displays are digital. In the 70s everyone wanted a digital watch; now we're another half-turn around the helix of fashion, and people want watches with hands.

Superficially, the distinction between digital and analog seems quite sharp, but this is really an illusion. There are, for instance, liquid-crystal-display watches whose on-off bits display a picture of watch hands! As I have already mentioned, people often imagine that the left half of the brain functions digitally, while the right half of the brain functions in an analog mode. As metaphor this is fine, but it is good to remember that the two brain halves are made of exactly the same kinds of components. In general, any digital process can be represented in analog form, and vice versa.

In practice, when we want to try and speak precisely about information it is better to take digital concepts as fundamental. The difference between digital and analog is really one of degree. The minimal comprehensible chunks of a digital communication require only a few information-code bits each, whereas the smallest meaningful pieces of an analog communication take quite a few bits apiece. A letter of the alphabet is an example of what I mean by a digital-like piece of info — in context, it takes only two or three bits to guess the right letter. A location of a line segment is an example of what I mean by an analog-like piece of info — a precision of one part in a thousand takes ten bits.

Once we realize that the difference between digital and analog communication is only one of degree, we can talk about analog-seeming communications without the fear of undermining our theory of information.

On an intellectual level, a written text also has an analog feel to it. One doesn't understand a page by decoding it one letter-bit at a time. The onrush of language sets up, if you will, sympathetic vibrations in the reader's brain, and this feels like an analog process. From a digital point of view, this means that the information is being transmitted so fast and so efficiently that there is no time to notice the individual bits.

The analog aspect of spoken communication is especially obvious. A person talking is nothing like a Morse-code beeper. Tone is important, as is stress, as are the lengths of the pauses between words. Just look at the changes that stress and pause can ring on a simple "I'm glad to see you":

I'm glad to see you.	(Even if no one else is.)
I'm *glad* to see you.	(What made you think I wouldn't be?)
I'm glad to *see* you.	(Instead of just talking on the phone.)
I'm glad to see *you.*	(But not the zhlubb you came with.)
I'm glad to see you.	(So stop asking if I am.)
I'm . . . glad-to-see-you.	(Are you glad to see me?)
I'm . . . glad . . . to . . . see . . . you.	(And I'm drunk.)
I'm . . . glad-to-see . . . *you.*	(Me Tarzan, you Jane.)

By using tone changes (rising and falling pitch) you can make a sentence into a question, an insult, or a throwaway. You can sing the sentence and sound festive or menacing, depending on the tune. At a higher level of complexity, you can say a sentence in some kind of put-on accent to add another twist to its meaning. Just think of meeting someone and saying "I'm glad to see you" in one of these various accents: E.T., Kermit, Santa, Hitler, Jagger, Reagan, Gomer Pyle.

Your personal communication is further enriched by body language — the faces you make, what you do with your hands, where you are looking, and so on. If you stare at your girlfriend's purse while saying, "I'm glad to see you," she knows you're broke. If you lick your lips while you're saying it, she knows she looks good enough to eat. If you smack your forehead, she knows you'd forgotten she was coming.

I could multiply examples endlessly. The analog resources of talk are so great that, in a pinch, an informative conversation can be carried out simply by repeating one or two words in various tones — recall Mike Nichols and Elaine May's classic "John and Martha" routine, where a man and a woman simply say each other's names over and over, sketching out an intricate play of rediscovery, reconciliation, and seduction.

Now, the fact is that all of these speech subtleties can be coded up as digital bits of information about loudness of voice, pitch of voice, location of hands, etc. Each of these quantities can be expressed as a number, which can be specified to as many bits of precision as is significant. In practice this is meaningless — you don't say "Hi" by handing someone a page full of zeros and ones — but in theory it is significant. The actual currency of our communication can always be thought of as made up of bits.

Let me stress, one last time, that to regard communication as an exchange of bits is not necessarily to say that bits are all there is. It is simply that this may prove to be a useful way of looking at the world. Bohr's philosophy of complementarity ensures that the opposite way of looking at the world may be just as useful. The two opposing viewpoints here might be called atomism and phenomenalism. The digital, atomistic approach breaks the world into bits. The analog, phenomenological view takes our knowledge of the world to be made up of continuously varying perceptions. The two world views are complementary; both are right. But now, for the purpose of better understanding the nature of information, we are going to think like digital atomists.

I now want to build up the notion that the information I can get about something in the world, say a person like you, is an endless string of zeros and ones. Let's look at some "digital" and "analog" questions I might ask you:

"Digital" questions	"Analog" questions
Are you from New York?	Where were you born?
Are you happy?	What kind of mood are you in?
Are you over 21?	How old are you?

Now, in reality, any answer you can actually give me is basically digital; any finitely long information I get from you can be coded as so-and-so-many digital bits. Thus, "Where were you born?" is usually answered by giving the name of the closest large city. If one supposes that there are one thousand large cities in the world, then answering this question takes ten bits of information. If I want to know which of the million largest cities we're near, I'm asking for some twenty bits of info. The more precision I want, the more bits I'm asking for. Usually, in conversation we don't want very many bits. Initially, when I ask, "What kind of mood are you in?" I'm only looking for one bit of info: good–bad; or it may be that I'm looking for two bits, a rating of your mood on the four-step scale A–B–C–D. If I ask, "How tall are you?" I do not expect you to give me the endless decimal expansion of some real number; I just want a four-bit answer that selects among the sixteen inch steps between 5′0″ and 6′4″.

As I get more and more information about a person *X*, I refine my picture of the person in two distinct ways: I add more detail to the answers of questions I've already asked, and I add new dimensions to the picture of *X* that I have. Asking "What month was *X* born in?" adds detail to my knowledge of *X*'s age, but asking "What is *X*'s favorite food?" adds an entirely new line of inquiry. On learning that *X*'s favorite food is #23 (shrimp), I can fill in detail by asking what size shrimp *X* prefers, and how they should be prepared.

An endless stream of data about a person or thing *X* can be thought of as tending, in the limit, toward a precise position in infinite-dimensional Hilbert space. Each axis or dimension corresponds to one particular line of inquiry. As the data comes in, we do not first get a precise position on one axis, and then a precise position on the next axis, and so on. We spiral around and around, in widening loops, always adding new topics and always returning to the old ones, but what is this "Hilbert space" I keep talking about?

Most readers will have heard of four-dimensional space. Physicists often speak of time as the fourth dimension, and of the world as a pattern in four-dimensional space–time. Spiritualists sometimes propose that ghosts live in a spatial fourth dimension, an occult direction leading out of our space. Mathematicians say that, just as the plane can be thought of as the set of all ordered pairs of numbers (x,y), four-dimensional space can be thought of as the set of all ordered

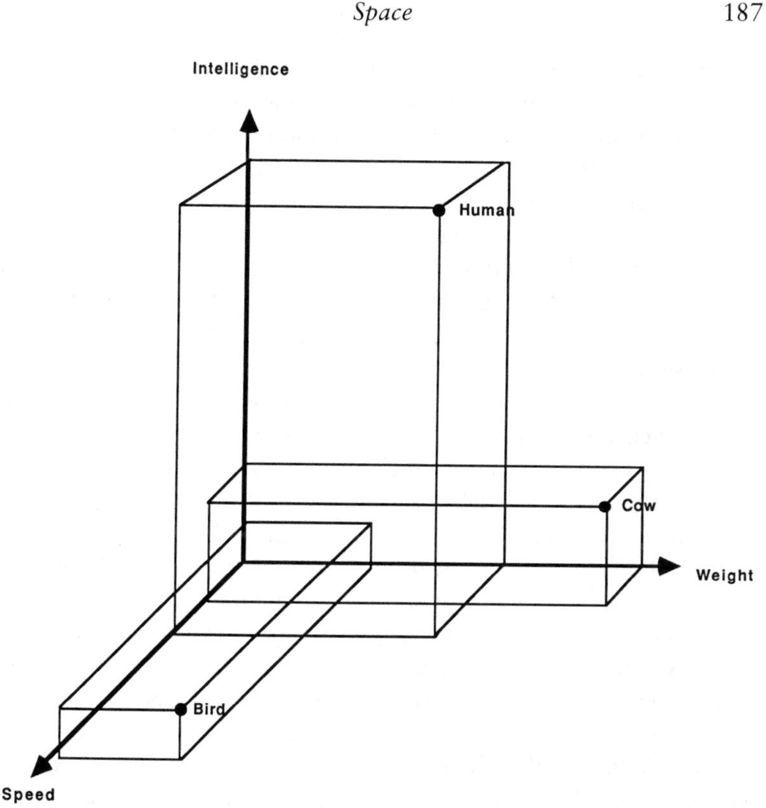

Fig. 105 A three-dimensional phase space.

quadruples (x,y,z,w). We are free to view the variable w as time, as a new direction, or as something else entirely.

For a mathematician, there is no great difficulty in going on to talk about spaces of five, ten, or a hundred dimensions. Physicists often use these higher-dimensional spaces as "phase spaces." If some individual object P has, say, seven distinct characteristics that we are interested in, then we might want to represent our knowledge of P as a region in a seven-dimensional phase space. Continuing in the same vein, there is no technical difficulty in talking about a space of endlessly many dimensions. This is what I mean by "Hilbert space." (The reader should be warned that books on quantum mechanics use

"Hilbert space" to stand for a specific kind of infinite-dimensional space whose coordinates are *complex* numbers.)

It can be argued that the whole question of the dimensionality of space is a bogus one. What we really have is an influx of information, and we arrange this information in certain ways. Certain kinds of arrangements are best thought of in spatial terms, but all that's really there is a lot of bits and a mnemonic system for remembering which bit means what. Mathematically, a real number x is thought of as having infinitely many digits in its full decimal expansion, so a one-dimensional number x is already as infinite as a location in Hilbert space. We shouldn't be scared to think of information as falling into multidimensional patterns.

Let me get a little more precise about the trade-off between increased precision and increased numbers of dimensions. To begin with, let's focus on the one-dimensional case. In its purest form, a stream of information can be thought of as an incoming flow of zeros and ones. A string of zeros and ones is usually called a "bit string." We could think of a bit string as being a base-two name for a whole number, but there is a disadvantage to doing so. The problem is that I can't begin interpreting a string of digits as a number until I have the whole string. In terms of ordinary decimal numbers, what I'm saying here is that if I tell you that a number starts with, say, 3, then you have no way of knowing if the number is in the 30s, the 300s, the 3000s, or what.

It's much better to interpret an incoming flow of zeros and ones as being the expansion of a binary point fraction. More and more digits of a point fraction determine a point x on the unit interval with greater and greater accuracy. Once I find out what range of sizes S I'm really interested in, I can view my bit string as the name of a number xS in this range. The virtue of this approach is that I can convert bits into location information as fast as they come in. In Table 2, we've shown what the various possible four-bit beginnings can mean, depending on what my scale range is.

Graphically, each new bit of precision halves the size of the interval of accuracy. As a practical matter, I can't draw on a piece of paper to more than six or seven bits of precision, but if I have a microscope, or if I am looking at a unit as long as a thousand pieces of paper, then I may be interested in dozens of bits of precision. For our purposes it is immaterial whether the lower end of a scale range is 0, 3, or 19.

Now suppose that we want to think of the bit string as being the name of a two-dimensional location. In the one-dimensional case, as we just saw, there is ambiguity about the scale on which to interpret a bit string. Turning the bit string into a location on a number line is a process of decoding. Each particular choice of scale and starting point decodes the bit string in a different way, but the basic notion of picking an interval and repeatedly halving it seems unique in its efficacy.

The best way of decoding a bit string into a two-dimensional lo-

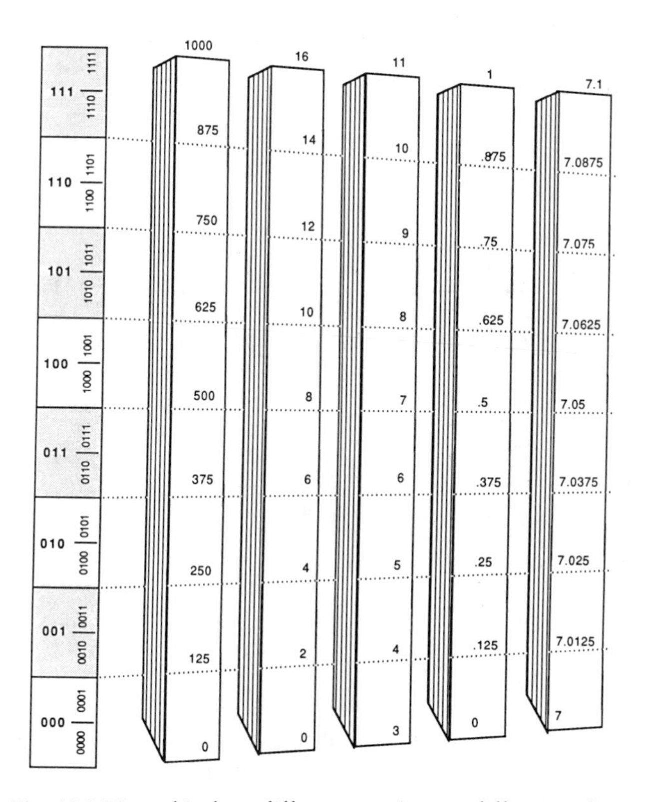

Fig. 106 Binary bits have different meanings on different scales.

Table 2.

Bit	*Range of values*				
code	*0–1000*	*0–16*	*3–11*	*0–1*	*19–19.86*
	0	0	3	.0	19
0000					
	62.5	1	.5	.0625	19.05375
0001					
	125	2	4	.125	19.1075
0010					
	187.5	3	4.5	.1875	19.16125
0011					
	250	4	5	.25	19.215
0100					
	312.5	5	5.5	.3125	19.26875
0101					
	375	6	6	.375	19.3225
0110					
	437.5	7	6.5	.4375	19.37625
0111					
	500	8	7	.5	19.43
1000					
	562.5	9	7.5	.5625	19.48375
1001					
	625	10	8	.625	19.5375
1010					
	687.5	11	8.5	.6875	19.59125
1011					
	750	12	9	.75	19.645
1100					
	812.5	13	9.5	.8125	19.69875
1101					
	875	14	10	.875	19.7525
1110					
	937.5	15	10.5	.9375	19.80625
1111					
	1000	16	11	1	19.86

cation also depends on the notion of repeatedly halving the possibilities. One good approach is to think of the bit string as being a shuffling of two separate, shorter bit strings. One of these strings gives

11	(00,11) 0101	(01,11) 0111	(10,11) 1101	(11,11) 1111
10	(00,10) 0100	(01,10) 0110	(10,10) 1100	(11,10) 1110
01	(00,01) 0001	(01,01) 0011	(10,01) 1001	(11,01) 1011
00	(00,00) 0000	(01,00) 0010	(10,00) 1000	(11,00) 1010
	00	01	10	11

Fig. 107 Shuffling x bits with y bits for two-dimensional names.

the location on the x axis, and the second gives the location on the y axis. Thus, relative to a unit square, we would interpret the bit string *0* 1 *1* 1 *0* 0 as specifying a location with x value *010*, and y value 110.

Looked at slightly differently, this approach regards the zeros and ones as a list of instructions. The instructions alternately choose between left or right and up or down. Thus 011100 becomes the list of instructions LURULD.

An equivalent (though different-seeming) way to do this is to look at the square diagonally and take the pairs 00, 10, 01, and 11 as basic symbols meaning "left, down, up, right." Then imagine subdividing the starting square into ever-smaller quadrants. Then 1011 means "Go to the bottom quadrant, and then go to the right subquadrant of that."

Something similar can be done for three-, four-, and seven-, or higher-dimensional spaces, though how one chooses to decode the string can vary on the amount of precision that one wants in each dimension. In general the cost of each new dimensional precision degree is one bit. Given six bits of information, I can view it as *XXXXXX,* a one-dimensional location known to six bits of precision; as *XYXYXY,* a two-dimensional location known to three bits in each direction; or as *XYZXYZ,* a three-dimensional location known to

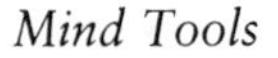

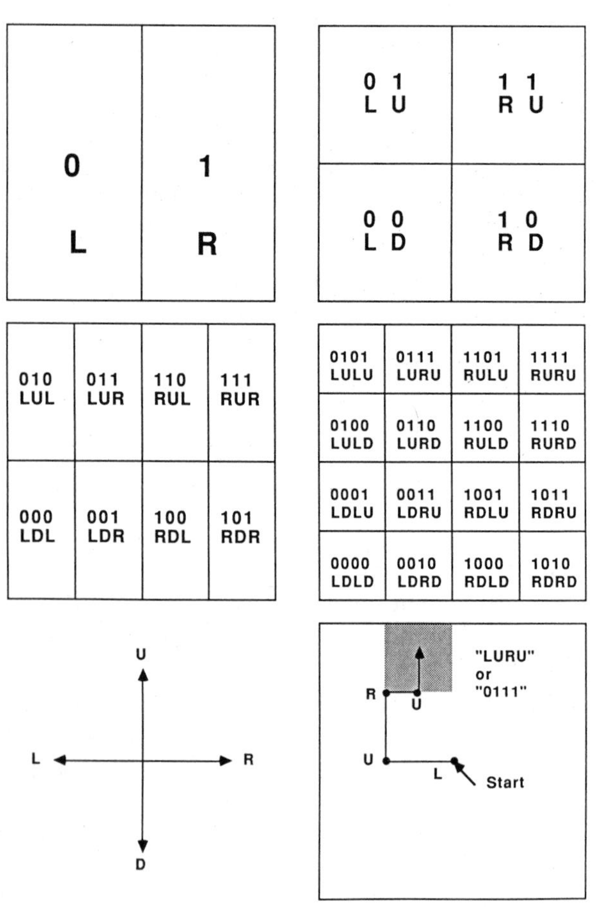

Fig. 108 A second way to get the same coding.

two bits in each direction. Using still another coding, we could think of the six bits as having the form *X–XY–XYZ*, giving the location to three bits in the *X* direction, to two bits in the *Y* direction, and to one bit in the *Z* direction.

If I have an endless bit string of information flowing in, then one way to think of it as giving an infinite amount of precision on each of infinitely many axes *A, B, C, D, E,* . . . is to imagine the code as spiraling ever outward:

A AB ABC ABCD ABCDE ABCDEF ABCDEFG ABCDEFGH . . . .

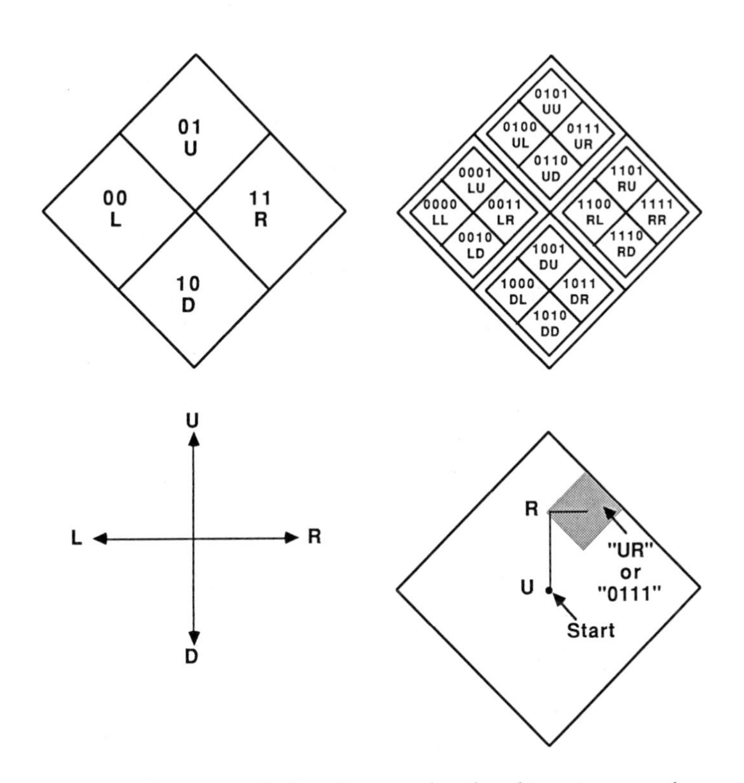

Fig. 109 Another way to look at how we decode a bit string as a plane location.

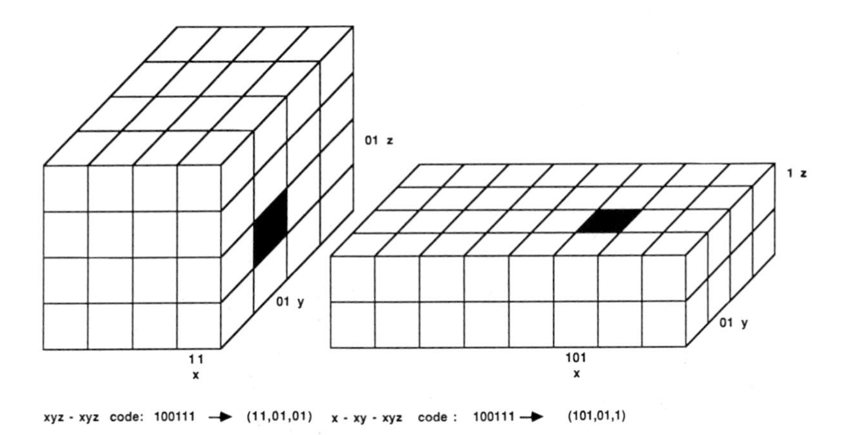

Fig. 110 Turning bit strings into three-dimensional space.

In the limit, one gets an endless number of bits for each of the positions *A, B, C, D* . . . . This decoding describes a position in a Hilbert space in which each axis represents an analog choice of a location on a line segment.

An alternate way of getting an infinite-dimensional location is to be content with one bit of precision on each "axis." In this case, we are just using the code form *ABCDEFGH* . . . . This decoding describes a position in a Hilbert space in which each axis represents a digital choice between zero and one.

3

LOGIC

The Laws of Thought

In one of the Marx Brothers movies, Chico and Zeppo get themselves into some terrible predicament. They're locked in a room and pacing desperately. "We've got to *think!*" cries Zeppo in his toffee-toned voice. Chico makes a dismissive hand gesture. "Nah. We already tried dat."

Much the same can be said about logic. On the face of it, the application of formal logic might be expected to resolve all kinds of disagreements. As it turns out, though, the known laws of logic are too few in number to be of any great help.

It is sometimes said that the history of logic breaks into three historical periods: the classical period, consisting of nearly two millennia dominated by the thought of Aristotle; the algebraic or symbolic period, running from Boole to Hilbert; and the modern or metamathematical period, running from Godel to the present. In the next three sections of this chapter we will discuss these three eras of logic, and in the last sections we will see how logic is now moving into a strange fourth era. In this introductory section, I would simply like to dwell on the gap between human hopes and human achievements in the realm of logic.

The seventeenth-century philosopher Gottfried Leibniz was one of logic's great dreamers. He contemplated the creation of a universal language, to be called the *characteristica universalis,* and a precise science of reasoning, the *calculus ratiocinator.* In Leibniz's vision, a day would come when disagreeing parties would be able to sit down together, take out pencil and paper, and say, "Let us calculate."

But how far did Leibniz actually get with his grandiose project? He managed to formulate the principle of noncontradiction, which tells us that no sentence is both true and false; and he formulated the principle of indiscernibles, which tells us that two things are equal exactly when all their properties are the same. Try and calculate all truth from that!

In 1854, the Irish mathematician George Boole carried out a concerted attempt to push Leibniz's program closer to conclusion. In his classic work, *The Laws of Thought,* Boole tried to treat logic as a simple, mechanical process akin to algebra.

Boole's book begins optimistically: "The design of the following treatise is to investigate the fundamental laws of those operations of the mind by which reasoning is performed; to give expression to them in the symbolical language of a Calculus, and upon this foundation to establish the science of Logic and construct its method." Boole describes his basic laws of logic and goes on to apply them to what must have seemed like the most important arguments of all: proofs by Clarke and Spinoza of God's existence. These analyses are somewhat inconclusive, and by the end of the book, Boole is in a state near despair: "Herein too may be felt the powerlessness of mere Logic, the insufficiency of the profoundest knowledge of the laws of the understanding, to resolve those problems which lie nearer to our hearts, as progressive years strip away from our life the illusions of its golden dawn."

As we will see, the actual rules of logic are pitifully few in number. It turns out that what is of real significance is the complexity of the initial assumptions from which you draw logical conclusions and the length of the logical arguments you are prepared to carry out.

Most computers have programs that generate "random numbers," but these numbers are not truly random; they result from hidden computations that are complex enough to not be easily predictable. By the same token, the fact that people often seem to act randomly does not imply that human nature is essentially illogical. It is simply that the logical processes underlying human behavior are too complicated to be subject to simple predictions.

Modern logical studies show that it is a mistake to expect a complex, long-running system to be simulated by any quick and easy logical rules. People who would like to think of themselves as rigorously logical are sometimes quite upset by the impossibility of really understanding the rules by which they operate. A perfect example of a person torn by the disparity between his inner complexity and his dream of logic is the leading nineteenth-century popularizer of symbolic logic, the Rev. Charles Lutwidge Dodgson (1832–1898), better known as Lewis Carroll.

As well as writing the classic fantasy *Alice's Adventures in Wonderland,* Carroll wrote a number of serious scholarly works. Toward the end of his life, most of his efforts were devoted to a two-volume work called *Symbolic Logic*. Only the first volume appeared in his lifetime. Appropriately enough, it is dedicated to the memory of Aristotle.

Although the whole world treasures his *Alice,* Carroll regarded his *Symbolic Logic* as his most important work. A lady named Edith Olivier, who often dined with him at Oxford, reports, "Mr. Dodgson never spoke of *Alice in Wonderland;* but there were three other things in his life of which he seemed really proud. He spoke of them every time we had dinner together. They were his kettle, his logic, and his photographs."

With this in mind, it is interesting to see the kind of self-image Carroll had the year he wrote the immortal *Alice*. The tale arose more or less spontaneously as a long story that Carroll told on July 4, 1862, while out rowing with the three little Liddell girls: Ina, Alice, and Edith. At Alice's repeated promptings, Carroll late in 1862 began writing the story down, and he finished the first draft during 1863. Here are two entries from *The Diaries of Lewis Carroll,* entries written, respectively, at the beginning and at the end of 1863, the year Carroll finished *Alice in Wonderland,* his greatest work.

> *Feb 6, 1863*. I will, before I sleep, record some plans for my future life. This year has given no promise as yet of being better than its predecessors: my habits of life need much amendment — and I am grievously neglecting means of grace. With God's help I desire to begin (1) daily reading and meditation on the Bible, (2) . . . .

> *Dec 31, 1863*. Here, at the close of another year, how much of neglect, carelessness, and sin have I to remember! I had hoped during the year, to have made a beginning in parochial work, to have thrown off habits of evil, to have advanced in my work at Christ Church. How little, next to nothing has been done of all this! Now I have a fresh year before me; once more let me set myself to do something worthy of life.

In 1863 Lewis Carroll generated the richest information of his life. It is precisely because *Alice* is so complex that the mental processes that produced it cannot be logically analyzed in any simple way. Unable to grasp the roots of his inspiration, Carroll felt himself to be acting ineffectively and at random.

There is less actual machinery to logic than one might expect. The laws of logic are like the rules that govern individual switches in a computer — rules such as "if you input two ON signals to an AND gate, the output is ON." In order for the computer to produce anything interesting, it needs to be given a complex information pattern to process, and it needs to be allowed to run for a long time. By the same token, unless one starts with an interesting set of assumptions and carries out very long arguments, a formalized logic system will not generate any surprising results.

What we want to do now is learn the little bit of logic that there is, see how logic can be used to code up formal systems, and determine how rich a system has to be to write something as good as *Alice*.

The Syllogism

Plato is the earliest figure associated with the concept of logical argument. His dialogues are filled with characters who claim to prove this and that. Plato himself did not entirely trust logical argument, and he sometimes introduced characters known as "sophists," who take a perverse delight in proving everything under the sun by means of verbal tricks. A well-known example occurs in the dialogue known as *Euthydemus,* where Socrates and his friend Ctesippus are arguing with a pair of sophists known as Euthydemus and Dionysodorus:

Dionysodorus: You say that you have a dog.
Ctesippus: Yes, a villain of one.
D: And he has puppies?
C: Yes, and they are very like himself.
D: And the dog is the father of them?
C: Yes, I certainly saw him and the mother of the puppies come together.

D: And is he not yours?
C: To be sure he is.
D: Then he is a father, and he is yours; ergo, he is your father, and the puppies are your brothers.

This kind of sophistical argument is based on a deliberate confusion about how to combine the separate properties of something being both "yours" and "a father." The argument works by departing from the accepted correct use of the word "yours." When Socrates tries to point this out to Dionysodorus, Dionysodorus tells Socrates, "You are prating, and are an ancient."

Plato's dialogues seem more approachable once you realize that lots of people in them talk to Socrates that way. The fact is that Plato and his alter ego Socrates never seem to be entirely sure they are right. The difficulty is that we really don't know how to formulate *all* the rules for correct language use. The formal logic that humans have developed deals only with the correct use of certain very special kinds of statements.

Nowadays most college mathematics courses begin with a few lectures on logic. Far from being a full introduction to the field, these lectures only concern the correct usage of the so-called logical connectives: not, or, if and only if. One learns facts such as: If A is true, and B is false, then (A implies B) is false. This whole subject area, technically known as *propositional calculus,* has been very well understood since the work of Boole and other mid-nineteenth-century logicians.

It often happens that a science does not historically begin by studying the simplest possible concepts. More often than not, the founders of a science are eager to get to work on the hardest puzzles the field has to offer. "Forget this 'crawl before I can walk' stuff. I want to fly!" So it is perhaps not so surprising to learn that Aristotle, the first real logician, did not concern himself very much with the simple rules of propositional calculus.

Instead of formulating rules about elementary combinations of true-or-false propositions, Aristotle spent the most time on what would now be called "the calculus of monadic properties." That is, he was interested in sentences that talk about properties that individual things might have.

The simplest sentences of this nature say something about two distinct properties. I am thinking of a sentence such as "Every dog is hairy." Put a bit more artificially, this sentence says "Everything which has the property of *being a dog* also has the property of *being hairy.*" If we think of property *S* (for subject) as standing for "dogness," and property *P* (for predicate) as standing for "hairiness," we see that our sentence has the form "Every *S* is *P*."

Now, according to whether we talk about every *S* or only some *S*, and according to whether we talk about things being *P* or not being *P*, we come up with four kinds of sentences about two properties:

Type	Specific example	General form	Name of the form
A	Every dog is hairy.	Every *S* is *P*.	Universal affirmation.
I	Some dogs are hairy.	Some *S* are *P*.	Particular affirmation.
E	No dog is hairy.	No *S* is *P*.	Universal negation.
O	Some dogs are not hairy.	Some *S* are not *P*.	Particular negation.

The medieval logicians, who were great students of Aristotle, gave these sentence types the names A, I, E, and O, apparently from the Latin *A*ff*I*rmo (I affirm) and n*E*g*O* (I deny).

Aristotle was interested in ways in which two sentences like this could be linked together to reach a single conclusion. Such an argument, with two premises and one conclusion, is called a "syllogism." Probably the best-known syllogism is the old chestnut

All men are mortal;	Every *M* is *P;*	
Socrates is a man.	Every *S* is *M*.	
Socrates is mortal.	Every *S* is *P*.	1:AAA

In analyzing a syllogism like this, we speak of the first two lines as "premises" and the third line as the "conclusion." The first property mentioned in the conclusion is called *S* (for subject) and the second conclusion mentioned in the conclusion is called *P* (for predicate). In a valid syllogism, there will always be a third property, called *M* (for middle term), which is mentioned in the two premises. In the syllogism above, "being Socrates" is the *S* property, "being mortal" is the *P* property, and "being a man" is the *M* property.

There are two ways of characterizing a syllogism. The first characterization has to do with the kinds of sentences used. In the Socrates syllogism above, all three sentences are the universal affirmative Type A sentences. The syllogism is thus said to be in the "mood" AAA. Other moods are EIO, AOO, etc. All in all, there are 4 × 4 × 4, or 64, moods.

The AAA mood is sometimes called Barbara: b *A* r b *A* r *A*. "Barbara" here is not the name of a girl, it is the plural of a Latin word for "barbarian." Even barbarians can put together a simple syllogism in the AAA mood.

The second way of characterizing a syllogism has to do with the arrangement of the properties *P*, *M*, and *S* in the two premises. There are four possibilities, known as "figures":

	First figure	Second figure	Third figure	Fourth figure
First premise	*M–P*	*P–M*	*M–P*	*P–M*
Second premise	*S–M*	*S–M*	*M–S*	*M–S*
Conclusion	*S–P*	*S–P*	*S–P*	*S–P*

All in all, since there are four possible figures and sixty-four possible moods, there are 4 × 64, or 256, possible syllogisms. Of these, nineteen represent correct modes of reasoning.

Listed below are examples of the nineteen correct syllogisms. In reading them keep in mind that, although the premises may or may not be true, the laws of syllogistic reasoning ensure in each case that *if* the premises are valid, *then* so is the conclusion.

Example	Form	
Brains wear glasses; Scientists are brains. Scientists wear glasses.	Every *M* is *P*; Every *S* is *M*. Every *S* is *P*.	1:AAA
No beggar is honest; All evangelists are beggars. No evangelist is honest.	No *M* is *P*; Every *S* is *M*. No *S* is *P*.	1:EAE
All men are beasts; Some saints are men. Some saints are beasts.	Every *M* is *P*; Some *S* are *M*. Some *S* are *P*.	1:AII
The only good writer is a dead writer; Some Americans are good writers. Some Americans are dead.	No *M* is *P*; Some *S* are *M*. Some *S* are not *P*.	1:EIO
No teachers are enthusiastic; You are enthusiastic. You are not a teacher.	No *P* is *M*; Every *S* is *M*. No *S* is *P*.	2:EAE
Every dog has his day; No skinflint has his day. No dog is a skinflint.	Every *P* is *M*; No *S* is *M*. No *S* is *P*.	2:AEE

No president is a moron;
Some illiterates are morons.
Some illiterates are not president.

No *P* is *M;*
Some *S* are *M.*
Some *S* are not *P.* — 2:EIO

Every good book is readable;
Some classics are not readable.
Some classics are not good books.

Every *P* is *M;*
Some *S* are not *M.*
Some *S* are not *P.* — 2:AOO

Every weekend I play golf;
Some weekends I'm with my father.
Sometimes I play golf with my father.

Every *M* is *P;*
Some *M* are *S.*
Some *S* are *P.* — 3:AII

No man is an island;
Some men float.
Some floating things are not islands.

No *M* is *P;*
Some *M* are *S.*
Some *S* are not *P.* — 3:EIO

Some days I'm happy;
Every day I have the blues.
Sometimes I'm happy when I have the blues.

Some *M* are *P;*
Every *M* is *S.*
Some *P* are *S.* — 3:IAI

Some women are not pretty;
Every woman is lovely.
Some lovelies are not pretty.

Some *M* are not *P;*
Every *M* is *S.*
Some *S* are not *P.* — 3:OAO

Every sex act is nasty;
Every sex act is sacred.
Some sacred things are nasty.

Every *M* is *P;*
Every *M* is *S.*
Some *S* are *P.* — 3:AAI

No parents love war;
All parents are producers.
Some producers do not love war.

No *M* is *P;*
Every *M* is *S.*
Some *S* are not *P.* — 3:EAO

Everything he likes is esoteric;
No esoteric things are on TV.
Nothing on TV is what he likes.

Every *P* is *M;*
No *M* is *S.*
No *S* is *P.* — 4:AEE

No criminal is kind;
Some kind people are poor.
Some poor people are not criminals.

No *P* is *M;*
Some *M* are *S.*
Some *S* are not *P.* — 4:EIO

Some important truths are obvious;
Every obvious thing is dull.
Some dull things are important truths.

Some *P* are *M;*
Every *M* is *S.*
Some *S* are *P.* — 4:IAI

Every person is an object;	Every P is M;	
Every object is a pattern in Hilbert space.	Every M is S.	
Some patterns in Hilbert space are people.	Some S are P.	4:AAI
Nothing I explain takes long to read;	No P are M;	
Things that take long to read are profound.	Every M is S.	
Some profound things are not explained by me.	Some S are not P.	4:EAO

The numbers and letters to the right of each syllogism have to do with the above-mentioned system for classifying them. In each case, the three letters state in order the types of the three sentences used; this, again, is the "mood" of the syllogism. The numbers give the "figure" of each syllogism. [Note: There are in fact five more valid syllogism forms. These are the forms 1:AAI, 1:EAO, 2:AEO, 2:EAO, and 4:AEO. Each of these forms is a weaker version of one of the principal nineteen valid forms.]

After his death in 322 B.C., Aristotle's logical writings were collected into a book known as the *Organon*, which means something like "instrument of science." Thousands and thousands of young men studied Aristotelian logic at the universities of the Middle Ages. One of the more popular texts was the *Introductiones in Logicam* of William of Shyreswood, published in Paris in the first half of the thirteenth century. An interesting feature of this book is that it includes a nonsensical mnemonic poem intended to help students remember the moods of the correct syllogisms in each figure:

> Barbara celarent darii ferio baralipton
> Celantes dabitis fapesmo frisesomorum;
> Cesare campestres festion baroco; darapti
> Felapton disamis datisi bocardo ferison.

If we only look at the vowels in the Latin words, we can view the first-line words, "Barbara celarent darii ferio," as a code for the four correct first-figure syllogism moods: AAA, EAE, AII, and EIO. Due to various constraints, the rest of the lines do not match up perfectly with the other three figures of the syllogism. One constraint is that consonants in William of Shyreswood's verse are of mnemonic significance. Another constraint is that the medieval view of what constitutes a correct Aristotelian syllogism differs a bit from the modern view.

Getting into the crazed, medieval spirit of all this, I've felt compelled to invent an English mnemonic verse called "Four Hours in a Pub." Each line is in two parts. The first part of each line represents the *figure* by an arrangement of the letters *P, M,* and *S.* The second part of each line represents the valid *moods* of the figure by a series of three-vowel words; the words with less than three vowels are only there to add shreds of meaning.

FOUR HOURS IN A PUB

The lamp smiles . . . My newspaper mentions raising bananas.
PM smites: The bleared barroom section agrees.
Map the mess: Roadhog detractors mistaking dadaist paintings for venison!
Drink Pimms . . . Inhaling the Creator's tension, careen to no avail.

Table 3.

Premises	*Form*
1. Abstract concepts are independent of the material world.	Every *A* is *B*.
2. Any information is an abstract concept.	Every *C* is *A*.
3. Nothing independent of the material world is mortal.	No *B* is *D*.
4. A brain's pattern is a kind of information.	Every *E* is *C*.
5. Your mind is a pattern in your brain.	Every *F* is *E*.
Conclusion: Your mind is immortal.	No *F* is *D*.

Steps of proof	*Reasons*
Every *A* is *B*.	Premise 1
Every *C* is *A*.	Premise 2
Every *C* is *B*.	Syllogism 1:AAA
No *B* is *D*.	Premise 3
No *D* is *C*.	Syllogism 4:AEE
Every *E* is *C*.	Premise 4
No *E* is *D*.	Syllogism 2:EAE
All *F* is *E*.	Premise 5
No *F* is *D*.	Syllogism 1:EAE

Note the verse's key features:

MP–SM	EAE EIO AII AAA
PM–SM	EAE AOO EIO AEE
MP–MS	OAO EAO IAI AAI AII EIO
PM–MS	IAI EAO EIO AEE AAI

One use for syllogisms is to link them together into long chains of reasoning. Such a chain of syllogisms is known as a *sorites,* from the Greek word for "heap," or "pile." Table 3 is an example of a sorites that takes five premises and proves that your mind is immortal.

As the syllogistic form of my sorites is correct, Aristotelian logic dictates that if the premises are all true, then the conclusion has to be true as well. Of course, if you reject some of the premises, then you are not compelled to accept my conclusion.

Symbolic Logic

Although Aristotle's method of syllogistic reasoning is certainly correct, many thinkers have felt that logic should be able to go beyond Aristotle's simple truths. Already in the second century A.D., the physician Galen pointed out that pure syllogisms are not adequate to handle arguments that involve relations between pairs of things. Two of Galen's examples are:

Mark has twice as much as Peter;	*M* is *K* to *P*;
Sarah has twice as much as Mark.	*S* is *K* to *M*.
Sarah has four times as much as Peter.	*S* is *KK* to *P*.

Plato is student of Socrates.	*P* is *J* to *S*.
Socrates is teacher of Plato.	*S* is J^* to *P*.

Many thinkers came to feel that the kinds of properties discussed by syllogisms are too restricted; there was a suspicion that there might be some methods of correct reasoning that Aristotle had overlooked. As the use of algebraic symbols and equations took over mathematics, people began to dream of an "algebra of logic" that would bring new power and simplicity to human reasoning.

The last big push to understand logic took place around the turn of the century. Logicians stopped endlessly rewording Aristotle's teachings and gave the whole subject a fresh examination. Some of the best-known logical explorers of this period (each listed with a key publication and its date) are

George Boole	*The Laws of Thought*	1854
Gottlob Frege	*Begriffschrift*	1879
Giuseppe Peano	*Notations de Logique*	1894
Bertrand Russell	*Principia Mathematica*	1910 (With A.N. Whitehead)
David Hilbert	*Uber das Unendliche*	1926

Basically, Boole got control over the logical connectives *or, and,* and *not;* Frege took care of *implies* and *for all;* Peano introduced a flexible symbolism; Russell showed how to put all of mathematics into a standard logical form; Hilbert showed how to build up logical systems of many different kinds. These men sent great bright streamers out from the known core of self-evident truths. Changing the metaphor, they helped the seed of truth to germinate, and to spread shining fractal vines far and wide.

Boole thought of logic in terms of algebra. He regarded a sentence as having value one if true, and value zero if false. For his arithmetic of truth values he makes the special assumption that $(1 + 1)$ is the same as one; the justification for this has to do with Boole's notion that zero is nothing, and one is everything. Keeping up the algebraic analogy, Boole thinks of "A and B" as being the same as "A times B," and he thinks of "not-A" as being the same as $1 - A$.

Under Boole's way of looking at things, various metaphysical principles become algebraic truths. Consider Leibniz's principle of noncontradiction, the statement that no sentence is both true and false. Put differently, the principle says: "(A and not-A) is false." Boole derives this principle from a simpler statement in an algebraic way.

To begin with, he adopts the assumption $A = A^2$ as a "fundamental law of thought," and then he argues

$A = A^2$
$A - A^2 = 0$
$A \times (1 - A) = 0$
(A and not-A) is false.

By making logic look like algebra, Boole hoped to give logic the rigor and inevitability that algebra enjoys. Yet, as Frege and Russell were later to show, algebra really presupposes a certain amount of logic, so it is better to build up your logic *before* you start talking about algebra. Another problem with Boole's approach is that his too-close analogy to algebra makes the notation hard to read. Peano gave logic a much nicer look by formulating all logical statements in terms of seven basic symbols:

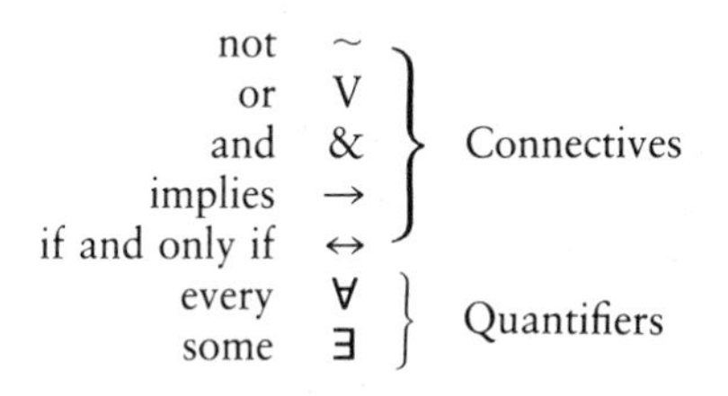

As indicated, the first five of these logical symbols are called "connectives," and the last two logical symbols are called "quantifiers." When we look at sentences that involve all seven of the basic logical symbols, we are studying what is known as "predicate calculus." Aristotelian logic is a part of predicate calculus. We can use the logical symbols A, I, E, and O to write down examples of the four basic kinds of sentences.

Let x be any object, let $D(x)$ mean "x is a dog," and let $H(x)$ mean "x is hairy."

TYPE	*ENGLISH FORM*	*SYMBOLIC FORM*
A	Every dog is hairy.	$(\forall x)[D(x) \rightarrow H(x)]$
I	Some dogs are hairy.	$(\exists x)[D(x) \,\&\, H(x)]$
E	No dog is hairy.	$(\forall x)[D(x) \rightarrow \sim H(x)]$
O	Some dogs are not hairy.	$(\exists x)[D(x) \,\&\, \sim H(x)]$

Much more complicated kinds of sentences can be written in logical form as well. Suppose, for instance, that we assume x and y are numbers, and let $L(x,y)$ mean x is less than y. Then the sentence "There is no largest number" can be restated as "For every number x, there is a number y such that x is less than y," which has the symbolic form $(\forall x)(\exists y)[L(x,y)]$.

Predicate calculus turns out to be quite rich and complex, but before going on with it, I'd like to go a step backward and restrict our attention to logical sentences that use connectives, but no quantifiers. The study of these simpler kinds of sentences is called "propositional

A	B	Not - A: $\sim A$	A or B: $A \vee B$	A and B: $A \& B$	If A, then B: $A \rightarrow B$	A if and only if B: $A \leftrightarrow B$
T	T	F	T	T	T	T
T	F	F	T	F	F	F
F	T	T	T	F	T	F
F	F	T	F	F	T	T

Fig. 111 A truth table for the propositional connectives.

calculus." Propositional calculus is an elementary branch of the full predicate calculus. Propositional calculus is very well understood.

The standard meaning of the propositional calculus symbols can best be given by a truth table such as Fig. 111. To understand the truth table, assume that A and B are two sentences, each of which is definitely true or definitely false. It is not hard to see that there are, all in all, four possible combinations of the truth or falsity of the two sentences. The truth table tells us that propositional calculus takes $(\sim A)$ to be true precisely when A is false. The logical "or" is taken in an inclusive sense; that is, $(A \vee B)$ is true if A is true, if B is true, or if both A and B are true. On the other hand, $(A \& B)$ is true only when both A and B are true. The statement $(A \rightarrow B)$ holds whenever B is at least as true as A is; that is, $(A \rightarrow B)$ is false only when A is true and B is not true. Finally, $(A \leftrightarrow B)$ means that either both A and B are true, or both A and B are false.

Two remarks on pronunciation: $A \rightarrow B$ can be pronounced either "A implies B," or "if A, then B"; "$A \leftrightarrow B$" can be pronounced either "A if and only if B," or "A is logically equivalent to B."

One of the failings of formal logic is that the implication sign $\rightarrow$ does not capture any kind of notion of cause and effect — $(A \rightarrow B)$ really just says the same thing as $(\sim A \vee B)$, so according to logic, such silly sentences as "If snow is black, then I can fly" are thought of as true. Unfortunately, no one has been able to come up with a better simple definition for implication than the one just given. In any

A	B	(A→ B)	↔	(~B	→	~A)
T	T	(T)	T	F	(T)	F
T	F	(F)	T	T	(F)	F
F	T	(T)	T	F	(T)	T
F	F	(T)	T	T	(T)	T

Fig. 112 Truth-table proof of the Law of Contraposition.

case, the rudimentary logical notion of implication has been quite effective for many kinds of logical analysis.

A logical combination of sentences that is always true, regardless of the truth or falsity of the constituent sentences, is known as a "tautology." In general, deciding if something is a tautology can be done in a simple and mechanical way by drawing up a truth table. Here are some well-known tautologies — compound statements that are regarded as true for any sentences *A*, *B*, and *C*. For each tautology we can give a little truth table to show it is always true:

Law of noncontradiction	$\sim (A \mathbin{\&} \sim A)$
Law of excluded middle	$A \vee \sim A$
Law of double negation	$\sim (\sim A) \rightarrow \mathrm{A}$
DeMorgan's laws	$\sim (A \vee B) \leftrightarrow (\sim A \mathbin{\&} \sim B)$
	$\sim (A \mathbin{\&} B) \leftrightarrow (\sim A \vee \sim B)$
Law of contraposition	$(A \rightarrow \mathrm{B}) \leftrightarrow (\sim B \rightarrow \sim A)$ (proved in Fig. 112)
Modus tollens	$(\sim B \mathbin{\&} (A \rightarrow B)) \rightarrow \sim A$
Consequentia mirabilis	$(\sim A \rightarrow A) \rightarrow A$
Reductio ad absurdum	$(\sim A \rightarrow (B \mathbin{\&} \sim B)) \rightarrow A$

Now look at some ordinary-language examples of these basic tautologies:

Law of noncontradiction: You can't be both happy and sad.

Law of excluded middle: You either believe in logic or you don't.

Law of double negation: If you ain't got nothing, you got something.

DeMorgan's laws: If a sentence is neither true nor false, then it's both false and true. If nobody's both male and female, then everyone's female or male.

Law of contraposition: If cleanliness leads to godliness, then devilishness leads to filth.

Modus tollens: If the world were logical, then the world would make sense, but the world doesn't make sense, so the world isn't logical.

Consequentia mirabilis: Even if we shouldn't philosophize, we should philosophize to explain why not, so in any case we should philosophize.

Reductio ad absurdum: If the world didn't exist, then this sentence wouldn't exist, but clearly this sentence does exist. Therefore the world exists.

One striking thing about these arguments is that not all of them seem entirely true or sensible. The basic problem is that our language is not based on solidly true-or-false sentences A, B, C, and so on. We may hope, though, that our reality is based on some underlying "atomic" properties that do obey such logical principles as the law of noncontradiction.

In any case, once the notion of tautology is precise, we can formulate the notion of "logical consequence." The sentence Z is a logical consequence of the sentences $A, B, C, \ldots, X, Y$, provided that the formula $((A \,\&\, B \,\&\, C \,\&\, \cdots \,\&\, X \,\&\, Y) \rightarrow Z)$ is a tautology. To say Z is a logical consequence of the sentences A through Y means that if A through Y are true, then Z has to be true.

A sentence that one chooses to believe in is usually known as an "axiom." If you lump together some of your favorite axioms, this collection of sentences comprises a "theory." If T is such a collection of axiomatic sentences, then we say that a sentence Z is "proved" by T if Z happens to be a logical consequence of some of the sentences in T, and Z is called a "theorem" of T. Symbolic logic allows us to think of a thoery T as compactly summarizing all the various sentences it is able to prove. In terms of information, the diverse consequences of a theory are coded up by the theory's axioms.

Exploring Logical Space

I think of "logical space" as being a big multidimensional space with each region corresponding to an assertion of some kind. Some assertions I know to be true. These areas are, let us say, brightly lit up. The assertions that I know to be false are darkly shadowed, and all the other regions are lit more or less brightly according to how likely I think they are.

Taking the space analogy to heart, I can compare my home planet to the many definite things that I know. "I am alive," "Grass is green," "My wife is female," "My dog is hairy," "Water is wet," "Ten is triangular" — these are all assertions that I know to be true. I live in the midst of many brightly shining truths. Hung in the distance like stars in the sky are certain other groups of truths. What I know about New York City is one cluster; what I know about relativity theory is another cluster; what I know about the natural number system is yet another distant gleam of truth.

Logic enables me to send luminous tendrils out from the facts I actually know. Seeing that my friend Rick's art shop is closed, I deduce that he has left on his badly needed Caribbean vacation. Seeing garbage cans by the curb, I deduce that the garbage truck is coming. Hearing a shrill cry from upstairs, I deduce that my children are fighting. These deductions are like simple syllogisms. We use them much more often than we realize. Sorites, or chains of syllogisms, are built into every aspect of our thought processes. Here is a sorites from Lewis Carroll's *Symbolic Logic:*

The only place to buy real *rubbish* is on the street.	Every *R* is *S*.
Nothing but rubbish can be had for a song;	Every *H* is *R*;
Things sold in the street are of no great value;	No *S* is *V*;
Eggs of the Great Auk are very valuable.	Every *E* is *V*;
An egg of the Great Auk is not to be had for a song.	No *E* is *H*.

This is a rather simple sorites (of type 1:AAA; 4:AEE; 1:EAE) with a rather obvious conclusion. When we look at mathematical deductions, however — the proof of the Pythagorean theorem, or the proof of the irrationality of the square root of two — we get a feeling that a complex-enough logic can extract much more from the evidence than one would ever have expected.

The dream of Leibniz was that, if only we could master the right language and the right laws of reasoning, it would be possible to move out from a few self-evident truths and light up all the truths there are. *Fiat lux!*

These days it's hard to imagine people thinking that way. Modern times have brought the breakdown of every kind of centralized authority — church, state, traditional science, received notions of beauty, and the like. In the late twentieth century, the notion of getting all truth for practically nothing seems as ludicrous as the search for the Holy Grail. What was it Chairman Mao said on one of his days of revolutionary good faith? *Let a thousand flowers bloom.* Let endlessly many logical stars spread out their rays!

But yet . . . it *would* be nice if everything in the world had a reason, if it all made some sense, if there really *were* a Secret of Life, a Grail, a Philosopher's Stone. As it turns out, there are logical theorems that set upper bounds to the complexity of the information that any given finite system can generate. It is, in effect, futile to hope for any simple solution to the world's complexity.

If we look at logical space from very close up, it is natural to wonder if all knowledge can be built up from simple true-or-false sentences. Although many normal sentences, such as "I am happy," can take on various shades of truth, it could be that these higher truths are built up from small, definite, "atomic" truths. Coming back to the first paragraph of this section, I am wondering if the variously shaded regions of logical space are made up of small black and white dots.

The viewpoint that regards the world as a collection of simple true-or-false sentences is in fact known as "logical atomism." The most elegant presentation of logical atomism occurs in the first, brief section of Ludwig Wittgenstein's 1921 classic, *Tractatus Logico-Philosophicus:*

1 The world is all that is the case.
1.1 The world is the totality of facts, not of things.

1.11 The world is determined by the facts, and by their being *all* the facts.
1.12 For the totality of facts determines what is the case, and also whatever is not the case.
1.13 The facts in logical space are the world.
1.2 The world divides into facts.
1.21 Each item can be the case or not the case while everything else remains the same.

Wittgenstein has a very definite picture of reality in mind. He is talking in terms of the same kind of logical space that we have been discussing, an imaginary space filled with possible assertions. Ultimately, the world is not made of objects, but rather of facts. If the facts are brightly lit and everything else is dark, then the entire pattern of logical space is given. This is all the world is, a pattern in logical space.

Some thinkers find Wittgenstein's theses 1.2 and 1.21 debatable. Thesis 1.2 suggests that facts are atomic; it gives the impression that once you get down to facts you are through dividing. Recall once again that "atom" comes from *a* + *tom,* meaning "no cut." An atom is something that can't be further divided. For Wittgenstein, a fact is something much simpler than a complex statement like "I am happy." It may be that Wittgenstein takes a fact to be something like "this atom is next to that atom," or "at time *t,* this particle has space coordinates (x,y,z)." So far as I know, he never gave a definite example of just what kind of thing *would* serve as a simple enough statement to be a primitive fact. Thesis 1.21 seems to confirm the notion that Wittgenstein means a fact to be indivisibly simple. If a basic "item" is made up of parts (as is the notion of my "happiness"), then changing that item's truth changes the truth of the parts.

The image that seems to emerge is that, for Wittgenstein, logical space is something like a graphic on a computer screen. There is a definite grid of minimal "items," or simplest possible affirmations, and each of these spots is either lit up (true), or dark (false). This is a kind of number-based way of looking at reality.

At the level of normal discourse, things are smeared and vague, but logical atomism holds out the hope that there is a deepest level at which reality resolves into hard facts. At this point in history we are no more able to judge the correctness of logical atomism than we are able to judge if matter can be resolved into ultimately indivisible particles, and it may be that these questions have no real answers.

In any case, we do know that the realm of numerical mathematics allows itself to be based on simple, atomic facts. Two numbers are equal or they aren't, and there's no problem with vagueness. Working with Alfred North Whitehead, Bertrand Russell made the first sustained effort to fill out the logical space of mathematics. Russell and Whitehead's mammoth undertaking was the three-volume *Principia Mathematica* of 1910–1913. The *Principia* takes the basic laws of the predicate calculus and tries to derive — practically from nothing — all of mathematics. One fairly practical reason for doing this was to put mathematics on a firm logical basis; a more utopian reason for the project was that it serves as a kind of practice run for a Leibnizian program of trying to deduce *all* truth from a few basic principles. The *Principia* was not entirely successful in its aims, but the work did show the great power of long, formalized logical arguments. Soon after its appearance, David Hilbert streamlined Russell and Whitehead's techniques to develop his concept of "formal systems."

Hilbert was perhaps the last of the great universal mathematicians. In the course of his career, he did important work in every area of mathematics: in number theory, geometry, logic, higher calculus, and applied mathematics. Most of his career was spent in Göttingen; drawn by Hilbert, mathematicians and physicists flocked there from every corner of the globe.

Hilbert felt that mathematics could be thought of as a process of making logical deductions from certain basic assumptions. Thanks in part to his work, all the basic rules of propositional and predicate calculus are now known. That is to say, mathematicians now know all the correct rules for putting together arguments based only on the meanings of the logical symbols $\sim$, $\vee$, &, $\rightarrow$, $\leftrightarrow$, $\forall$, and $\exists$.

Hilbert's conception of mathematics as a formal system was based on the idea that all the familiar mathematical relations, such as $(x = y)$, $(x = 0)$, $(z = x + y)$, can be regarded as special predicates in the language of predicate calculus: $E(x,y)$, $Z(x)$, $P(x,y,z)$. In view of this, the axioms of mathematics can be thought of as sentences in the predicate calculus, and the process of proving theorems can be thought of as a simple matter of applying the known rules of logic to the various axioms.

We now think of a Hilbert-style formal system as being like a machine for proving theorems. The axioms of a formal system correspond to the machine's initial program, and the rules of logic correspond to the machine's rules of operation. Once the program is selected, the machine sets about proving theorems quite automatically. The hope is that mathematicians can find a program powerful enough to fill up all the regions of logical space that are of interest to them.

Hilbert's work makes it clear that it is very important to find a good set of axioms for our mathematics. Suppose we use the letter M to stand for the collection of basic mathematical assumptions we start with. This collection might include, for instance, all the things about math that people are taught in high school: $1 + 1 = 2$; $x + y = y + x$; two points determine a straight line; the square on the hypotenuse is the sum of the squares on the two sides; the roots of a quadratic equation can be gotten by the quadratic formula; the derivative of x-squared is $2x$; etc. This approach is very inefficient — M can be made much smaller — but the basic idea is that M is supposed to be rich enough to sum up everything we know about math. Fine.

Hilbert pointed out that a good set of axioms for mathematics should be "consistent." To say M is consistent means that if we start using the rules of logic to deduce things from the axioms in M, we will never run into a contradiction; M will not force us to try to make some region of logical space be simultaneously black and white.

Although Hilbert did not immediately realize it, another thing that we want from a good set of axioms is that the set of axioms be "complete." To say M is complete means that, given any hypothesis H about mathematics, it is possible to either prove or disprove H on the basis of the axioms in M. As the consequences of M spread out through logical space, every region becomes definitely white or definitely black.

Coming back to Wittgenstein for a moment, we can see that he felt the collection F of all the facts that "are the case" makes up a complete and consistent description of all logical space and thus, of all the world. Of course, the full collection F would be so inconceivably vast as to be wholly unmanageable; Hilbert hoped that, at least for the domain of mathematics, a reasonably compact system M could be found that would also be consistent and complete.

As we will see next, Hilbert's dream was doomed to fail.

Godel's Theorem

Looked at objectively, the world is much more complicated than a person is, but we have this persistent urge to *understand* the world, to find some way of summing it up in a simple way. A good scientific theory has the property of compressing a great number of facts into a few basic laws. Thus, for instance, one can start with Newton's three laws of motion and his law of gravitation, and predict the dates of the next hundred solar eclipses. Given Bohr's model of the atom, one can deduce which kinds of chemical compounds can be formed easily. Given some good axioms for mathematics, one can prove a wide range of true facts about numbers.

The great dream of rationalism has always been to find some ultimate theory that can explain *everything*. As a kind of warm-up for the final assault on Truth, logicians such as David Hilbert hoped to find a theory *M* that would compactly sum up all of mathematics. Mathematics here serves as a kind of "toy universe." If we could figure out how to get a final theory of mathematics, then we would be a step closer to getting a final theory of the real world.

An ideal theory *M* of mathematics should have three properties: *M* should be finitely describable, consistent, and complete. To say that *M* should be "finitely describable" means that it should be possible to write a finitely long book that clearly explains what it means to say "*S* is provable from *M*." Given a proposed proof of sentence *S*, it should be possible to look at the description of *M* and decide in a finite amount of time whether or not the proof follows the rules of the theory *M*.

The reason for this condition is that you don't really have a theory at all unless you can clearly tell other people what the theory is. If it weren't for this restriction, we could just say, "Let *Tr* be the collection of all true sentences about math." The problem with *Tr* is that if I try to tell what *Tr* is by listing all the true sentences, then it will take me forever. Though *Tr* is consistent and complete, it is not finitely describable in any obvious way. There is a similar problem if I say,

"Let *Hu* be the collection of all the mathematical statements that human beings will ever prove to be true." Since we have no way of knowing what kinds of proof methods may be used in the future, we cannot now give any finite description of how to tell whether a sentence *S* follows from *Hu* or not.

Quite formally, to say that a theory *M* is finitely describable means that it is possible to program a digital computer to print out in succession all the sentences *M* can prove. The idea is that one hopes to combine the finite program for *M* with increasingly large amounts of time to produce, in the limit, all true sentences. Thinking of a good system *M* is the human part, and the working out of all consequences is the mechanical part. As more and more time goes by, some rather surprising results may begin to appear; there is a sense in which runtime itself supplies information.

The second and third properties desired of a good theory *M* are consistency and completeness. If we think of *M* as being a machine that prints out sentences, then to say that *M* is consistent is to say that *M* will never print some sentence *S* and then later turn around and print out ~ *S* as well. To say that the theorem-printing machine *M* is complete is to say that, for any given sentence *S*, *M* will eventually print out either *S* or ~ *S*.

If a theory is not consistent, then it is going to prove and disprove all kinds of things without telling us which things are really true. A simple example of an inconsistent theory arises if I say, "Let *Ev* be the collection of all mathematical sentences." *Ev* is finitely describable, since I can in fact give a finite set of rules for what kinds of symbol strings are mathematical sentences. And *Ev* is complete, since for any sentence *S*, either *S* or ~ *S* follows from *Ev* (in fact *both* do). But *Ev* is inconsistent and basically worthless.

If a theory *M* is not complete, then there will be some sentences *S* that the theory does not decide. Given the perverse nature of the world, it is quite likely that some of the sentences undecided by *M* are going to be sentences that I'm really curious about. If I say, "Let *Gr* be all the mathematical sentences that are taught in grade school," then *Gr* is finitely describable (by the textbooks used in grade school), and *Gr* is consistent, but *Gr* is not likely to be complete. *Gr* will, for instance, not prove things that involve concepts like calculus and imaginary numbers.

* * *

Mathematics seems quite crystalline and simple. Logic seems very powerful. In the 1920s, people like Hilbert really did expect that there should be some ultimate theory *M* for mathematics that should be finitely describable, consistent, and complete. In the summer of 1930, though, a twenty-four-year-old graduate student named Kurt Godel showed that there is no such theory.

GODEL'S INCOMPLETENESS THEOREM

No theory of all mathematics is
- i. finitely describable, and
- ii. consistent, and
- iii. complete.

Let's look at Godel's theorem five different ways, from the standpoints of number, space, logic, infinity, and information.

Godel's Theorem from a Number Standpoint

Let *M* be a consistent theory of mathematics, and assume that *M* is finitely describable. Say, for instance, that there is a fat book that describes all the axioms and rules of inference *M* uses. Now, as we discussed in chapter 1, this book can be thought of as an "L27 string" that names a certain natural number that we may as well call #*M*.

The actual proof of Godel's theorem shows that there is a number sentence *P*(#*M*) about the number #*M* such that *P*(#*M*) is in fact true, but the theory *M* is unable to actually prove the sentence *P*(#*M*).

P(#*M*) says something like "there are no whole numbers that satisfy such-and-such an equation involving #*M*." In fact, there *aren't* any such numbers, but *M* can't prove this.

In other words, Godel's theorem shows that there is a concrete fact about whole numbers that *M* is unable to prove; this implies that *M* is not a complete description of the mathematical universe.

Godel's Theorem from a Space Standpoint

Once again, let *M* be a consistent theory of mathematics with a finite description. Going back to our notion of logical space, we can think of *M* as being a kind of seed in the logical space. Imagine that *M* is white, and that the undetermined parts of logical space are gray. Now think of *M* as proving certain things and disproving certain other things. The things that *M* is proving lie along white branches growing out of the seed, and the things that *M* is disproving lie along black branches growing out of the seed.

Godel's theorem says that no matter how trickily *M* is chosen, there are always going to be some gray areas that never get turned white or black. The proof itself uses our finite description of *M* to build up a set of instructions for finding a specific location in logical space that will stay gray.

Putting it a bit differently, it is possible that Godel's theorem can be viewed as a statement about tessellations. I think one could come up with a (possibly higher-dimensional) space and a collection of shapes that would represent the notion of provability. In this situation, a theory would correspond to a certain initial pattern of tiles. Godel's theorem would say that no initial pattern can be extended to a tessellation of the whole space. More precisely, Godel's theorem would show how to look at any starting pattern and find a region of the space that cannot be consistently tessellated out to.

Godel's Theorem from an Infinity Standpoint

Godel once told the philosopher of science Hao Wang that the original inspiration for his incompleteness theorem was his realization that "truth" has no finite description. Once he realized that the set *Tr* of all true sentences has no finite description, he could be sure that no finitely describable *M* can sum up everything in *Tr*.

How did Godel come to see that "truth" has no finite description? By thinking about the ancient Paradox of the Liar. This paradox, which was known even to Aristotle, consists of a single sentence *L*, "This sentence is not true." If *L* is true, then *L* is not true; and if *L* is not true, then *L* is true. Thus, *L* is both true and not true. This state of affairs goes against a very fundamental logical assumption, the law of noncontradiction (a sentence *can't* be both true and not true). We refuse to abandon the law of noncontradiction, so it must be that *L* is, for some reason, not really a sentence.

There are two odd things about L. The first odd thing is that it is a sentence that refers to itself: L says, "L is not true." The second odd thing about L is that it talks about the concept of "truth." Now, Godel was able to show that self-reference is not really such a big problem. By looking at sentences as L27 codes, we can think of sentences as being certain kinds of numbers, and we can find a mathematical way of representing self-reference. It turns out that if Q is any property of sentence code numbers, there is a tricky way of constructing a mathematical sentence that says, "The code number of this sentence does not have property Q." We describe the trick in the next subsection.

So it must be the second odd thing about L that is causing the trouble. It must be that there is no finite mathematical way to express the property of "truth," or "x is the code number of a true sentence."

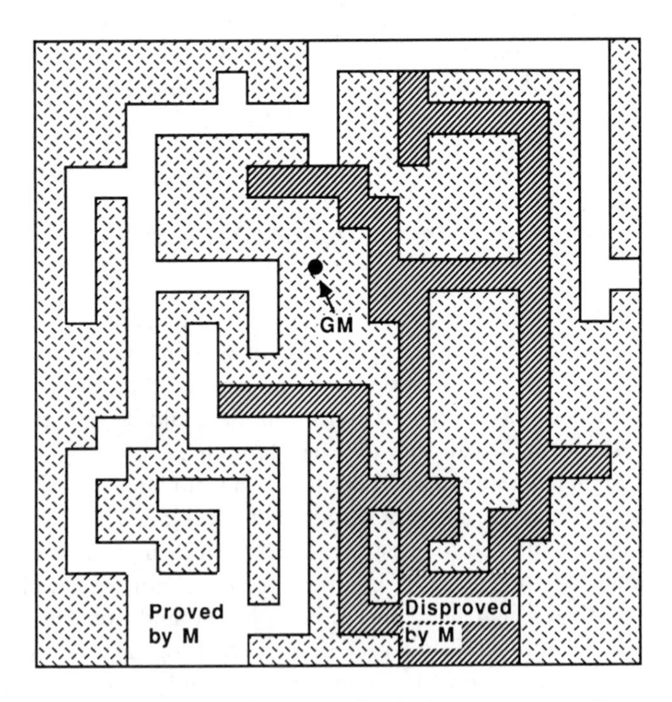

Fig. 113 Godel's theorem: "There's always a gray area."

This is how Godel came to see that "truth" has no finite mathematical description.

It is strange to realize that we can talk about a concept — truth — that is infinitely complicated, or maybe it isn't so strange. We talk about "the universe," even though we have no simple way of summing it all up. What is the universe? It's . . . *this.*

Your ability to understand what I mean by "this" hinges on the fact that you and I are similar beings who actually and miraculously exist. "This" is not really a rational concept; it's something so immediate that even dogs and rocks know about it.

It's not surprising that early logicians hoped to eliminate such a mystical residue from mathematical thought. Reasoning was what they were good at, and they would have liked to turn the whole world into pure reason. But Godel has shown that the fundamental logical notion of "truth" has no rational definition.

What is truth? *This.*

Godel's Theorem from a Logic Standpoint

Godel's incompleteness theorem does more than show that truth has no finite description. The proof of the theorem shows how to take any consistent and finitely describable theory *M* and construct a specific sentence *GM* which is true, but not provable by *M*.

GM is a sentence that says, "This sentence is not provable by the theory *M*." Although it doesn't look like it, *GM* really is a mathematical sentence, equivalent to the number sentence *P*(#*M*) discussed above. How is the sentence *GM* actually constructed? The construction proceeds in several stages.

Stage 1. We set up a coding process — such as the L27 code — by which any string *S* of mathematical symbols can be represented as a code number #*S*. This is pretty straightforward.

Stage 2. Since *M* is finitely describable, it is possible to find a finite mathematical property *Pf*(x), involving a variable x, that says, "x is the code number of a sentence provable from the theory *M*." This is done by letting *Pf*(x) say something like "there is a long number that codes up a correct proof of the sentence with code x from the axioms of the system M." The upshot is that if S is a sentence, then $Pf(\#S)$ means that S is provable from M.

Stage 3. This is the technical trick that makes self-referential mathematical formulas possible. We define a map that starts with a code number k and gives us another number called Sub(k). Suppose that k is the code number of a formula $K(x)$ that expresses a mathematical property of some variable number x. Thus K might be "x is even," or "x is a prime number," or "x is greater than 10,000." If we plug some specific number, such as 17, in for x, we get a specific sentence, which is written $K(17)$. $K(17)$ might be "17 is even," or "17 is a prime number," or "17 is greater than 10,000." Fine. Now recall that we started with a number k, which was the code number of K. Put differently, we started with the number k that is equal to $\#K(x)$. Now we can plug the number k into K, and get a mildly self-referential sentence $K(k)$. This sentence has its own code number, $\#K(k)$, and it is this number we are going to call Sub(k). In general, *if k is the code number of K(x), then Sub(k) is #K(k), the code number of K(k).*

Stage 4. Recall from Stage 2 that $Pf(x)$ is a mathematical formula that says, "x is not the code number of a sentence provable from the theory M." Now let $E(x)$ be the property "$\sim Pf(\text{Sub}(x))$," and then let e be the code number of $E(x)$. To finish things off, let GM be $E(e)$.

$GM \leftrightarrow E(e)$, by the definition of GM.
  $\leftrightarrow \sim Pf\,(\text{Sub}(e))$, because $E(x)$ is the property $\sim Pf(\text{Sub}(x))$.
  $\leftrightarrow \sim Pf(\#E(e))$, by the definition of Sub.
  $\leftrightarrow \#E(e)$ is not the code number of a sentence provable from the theory M, by the definition of Pf.
  $\leftrightarrow \#GM$ is not the code number of a sentence provable from the theory M, by the definition of GM.
  $\leftrightarrow GM$ is not provable from the theory M.

The point of all this is that GM is constructed so that if it is true, then it is not provable by M, and if it is false, then it *is* provable by M. Since M is supposed to be a good theory of mathematics, however, we don't want to have M proving anything that is false. Therefore the possibility that GM is false and provable can be ruled out, so it must be that GM is both true and unprovable. "This sentence is not provable by M" is a true sentence that M cannot prove.

Note that if there were a formula $Tr(x)$ expressing the property "x

is the code number of a true sentence," then we could carry out the same four-stage process to get a sentence L saying "This sentence is not true." Since L is self-contradictory, we see that, as was discussed in the last subsection, it must be that truth cannot be finitely expressed.

Godel's Theorem from an Information Standpoint

The IBM information theorist Gregory Chaitin, in the *International Journal of Theoretical Physics,* has summed this notion up very succinctly: "You can't prove a twenty-pound theorem with a ten-pound theory." The basic notion is that any given mathematical theory M incorporates a definite amount $I(M)$ of information. For an interesting theory M, the information $I(M)$ contained in M is roughly equal to the number of symbols it takes to write out a concise description of M. A more complicated theory codes up more information, but any given theory incorporates only a limited amount of information, and there are, way out there, numbers that code up much larger amounts of info. The information-theoretic version of Godel's theorem shows that M is unable to really talk about the numbers whose information-theoretic complexity is much greater than $I(M)$. The core insight, "our theories are finite, but truth is infinite," is sharpened by this result to something like, "for any of our theories M, there is an upper bound to the complexity of the truths that M can prove." Any notion of finding a theory that answers everything is abandoned here; theories are viewed as finite artifacts whose strength can be measured in an exact way. I'll say more about this version of Godel's theorem in the next chapter.

Godel's theorem has the monumental intricacy of a painting by Brueghel. There are many different ways to look at it. In the 1930s logicians used the techniques of Godel's theorem to arrive at a bewilderingly diverse collection of related results. Assisted by J. Barkley Rosser, Godel himself showed that his theorem means that no good mathematical theory can prove its own consistency. Alonzo Church used Godel's methods to show that there is no effective way to decide which sentences follow from a given theory. Alfred Tarski used Godel's proof to demonstrate the undefinability of truth. Alan Turing came up with a proof that no one can predict whether or not a given computer program will work. Emil Post discussed the way in which Godel's theorem shows that mathematics must be essentially creative and nonmechanical. Stephen Kleene related Godel's proof to the impossibility of finding names for every level of infinity.

Above all, Godel's theorem shows that human thought is more complex and less mechanical than anyone had ever believed, but after the initial flurry of excitement in the 1930s, the result ossified into a piece of technical mathematics. Godel's theorem became the private property of the mathematical logic establishment, and many of these academics were contemptuous of any suggestion that the theorem could have something to do with the real world. Throughout the materialistic, technology-worshipping 1940s and 1950s, the philosophical consequences of Godel's theorem were all but ignored. Interest in the theorem was rekindled in 1961 when the Oxford philosopher J. Anthony Lucas wrote an incorrect but very important paper arguing that Godel's theorem proves that machines cannot think. In recent years, Douglas Hofstadter's popular science writings have done much to increase the public awareness of Godel's theorem.

Buckminster Fuller once said that if an idea is really important, then it can be used to build better machines. In the next two sections, we'll look at some of the ways in which the techniques behind Godel's theorem are influencing modern thinking about computers.

Turing Machines

What does a digital computer actually do? A computer has four main components: input, memory, processor, and output. In a typical home computer, the input is a keyboard and a floppy disk, the memory is a grid of electronic on-off switches, the processor is a chip of integrated circuitry, and the output is a floppy disk, a video terminal, and a printer. The input is coded up in the memory, the processor alters what is in the memory, and then the output decodes and displays the new contents of the memory.

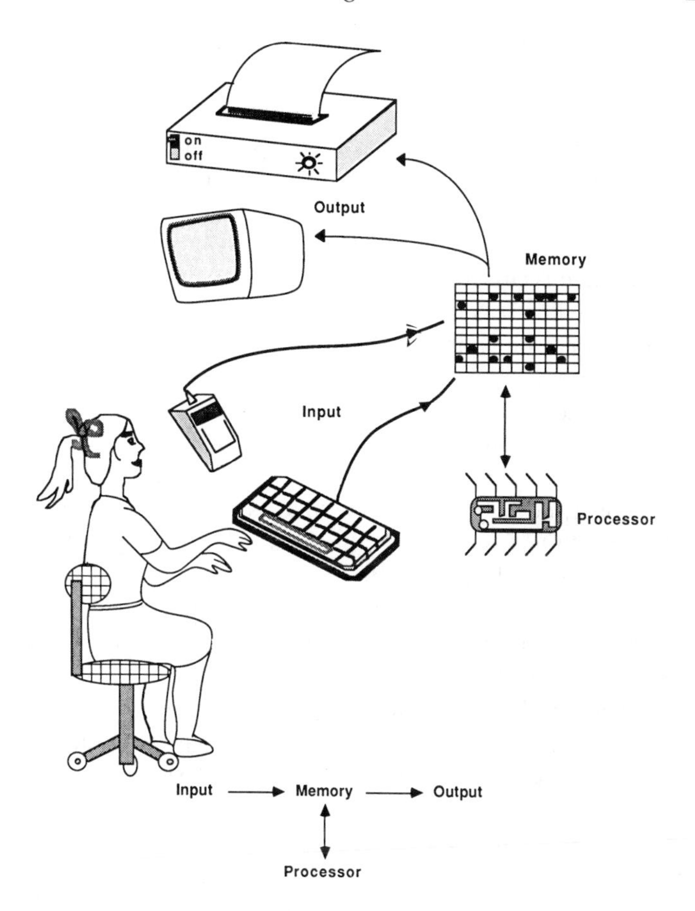

Fig. 114 A computer.

The actual computation consists of the processor's activities on the memory. Home computers have processors that are flat, bug-sized rectangles with lots of wires sticking out like legs. We might visualize a computer's operation as follows. Input stage: You insert program and data diskettes, and you type some things on your keyboard. A standard coding system turns all of this information into a pattern of zeros and ones on the memory board. Computation stage: Like a little bug, the chip "crawls around" on the memory board, repeatedly changing symbols. Output stage: The pattern on the memory board is decoded into marks on your disk, screen, and printer paper. Now exactly what is it that the processor has done?

i. The processor is able to read the symbol stored in whatever location it is looking at; that is, it can tell if the switch is set to on or off.
ii. The processor is able to change the contents of the memory location it is currently scanning; that is, it can change the position of the switch it is observing.
iii. The processor is able to move its attention to a new memory location.
iv. The processor is able to change its internal state; that is, it can change its predilections about what to do next.

This concept of what a computer does is drawn from Alan Turing's 1936 paper "On Computable Numbers, with an Application to the Entscheidungsproblem." His discussion of computation is based on a special, simplified computer called a "Turing machine." In principle, a Turing machine can carry out any computation that any other computer can do. In practice, they operate quite slowly, so no one ever actually builds them. Despite this, Turing machines are so well known that some well-meaning functionary has set aside a Library of Congress category number for books on "Marketing of Turing machines"!

A Turing machine looks something like a box with a long tape of paper running through it. The box also has a numbered stick, which can move back and forth. The numbers on the stick keep track of the machine's "internal state"; the length of the stick is analogous to the complexity of a computer's processor design. The paper tape is divided into square cells, and each cell is either blank or marked with a stroke /. The tape plays a role like that of the memory board above. To give the machine an input, you mark some of the squares and move the leftmost marked square to the machine's reading head. Now you turn the machine on by setting its state to 1. As time goes by, the machine moves the tape back and forth, one square at a time, occasionally putting down new /'s or erasing old /'s. The state bar jiggles back and forth. When the machine finishes its calculation, it stops moving and turns itself off. The output is the pattern of /'s remaining on the tape.

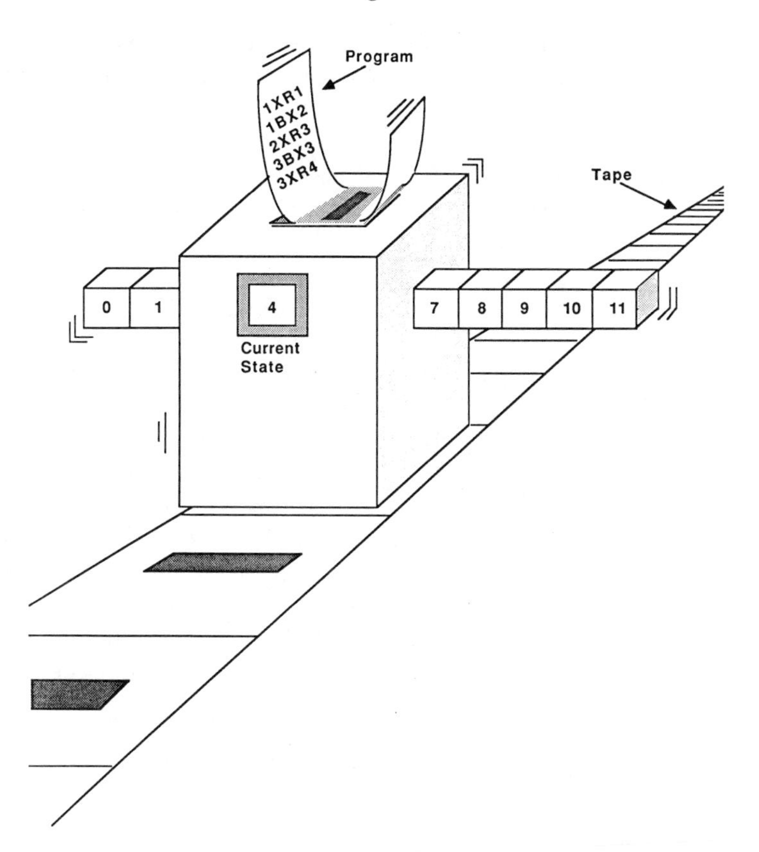

Fig. 115 A twelve-state Turing machine.

How does the machine decide which way to move the tape, how to change its state, when to erase, and when to make a mark? By means of its program. The program for a Turing machine is very simple — a finite list of four-symbol instructions. Each instruction consists of:

a) A natural number 1,2,3,. . . . These numbers stand for internal states that the machine might be in when it is looking at an instruction.
b) One of the letters *B* or *X*. These stand for descriptions of what the currently scanned cell looks like. *B* means blank, and *X* means marked.

c) One of the letters *B, X, R,* or *L*. These stand for actions the machine can take. *B* means "erase the mark in this cell"; *X* means "put a mark in this cell"; *R* means "move the tape one cell to the right"; and *L* means "move the tape one cell to the left."
d) A natural number 0,1,2,3,. . . . These numbers stand for the internal state the machine is in after completing the instruction. State 0 is the "turned off" state.

Typical Turing machine instructions are 1*XB*2, 2*BR*3, 3*BR*4, 4*BX*4, 4*XR*5, and so on. The instruction 4*XR*5 says, "If you are in state 4, and you are scanning a marked cell, then move one cell to the right and enter state 5."

We think of a Turing machine as being in state 0 when it's turned off, and as going into state 1 as soon as you turn it on by pushing the state bar over from 0 to 1. At that time the machine examines the cell it is scanning. If the cell is blank, the machine checks its list of instructions for an instruction that begins 1*B*. If the cell is marked, the machine checks its list of instructions for an instruction that begins 1*X*. Suppose it finds the instruction 1*XB*2. This says, "If you are in state 1, and if you are scanning a marked cell, then erase the mark and enter state 2." The machine continues in this way.

If and when the machine reaches an instruction that tells it to enter state 0, it turns itself off. Not every machine reaches an instruction that leads to 0; some machines go into various sorts of endless behavior loops. It is not at all unusual for a Turing machine to run forever. Turing's theorem says that there is no general method for distinguishing the machines that will run forever from the machines that will calculate a result and turn themselves off.

Keep in mind that any given Turing machine is capable of entering only a finite number of states. This is why it is useful to think of the machine as including a numbered bar that it slides back and forth to keep track of what state it is in, and to remember that the bar is only finitely long. As a rule, the more states a Turing machine has, the more complicated its behavior can be.

It is worth noting here that if a machine's number of nonzero states is fixed at the value *K*, then the machine's total program contains at most 2*K* instructions, each of which begins with one of the 2*K* possible prefixes 1*B*——, 1*X*——, 2*B*——, 2*X*——, · · · , *KB*——, *KX*——. This follows from the fact that any time the machine is on, its state is one of the numbers 1 thru K, and the square it is examining is either a B or an X. Each "––" can be filled in $4(K+1)$ ways; that is, by an L, R, B, or X, followed by one of the numbers 0 through K. This makes for $(4(K+1))^{2K}$ possible programs in all. This is quite a large number even for small Ks — for $K=2$ it is already about twenty thousand — but the number of possible K-state Turing machines *is* finite, and for large K, the value is roughly the same size as K^{2K}.

What can Turing machines be used for? Just about anything. The marks on a Turing machine's tape can stand for all kinds of different things, depending on what you are using the machine to calculate. This is analogous to thinking of the patterns of on-off switches on a computer's memory board as coding up various sorts of information, depending on the type of input–output devices being used. The memory bits can stand for numbers, letters, pixels on a screen, or settings on a rocket's controls.

For now, let's think of our Turing machine as being used to convert whole numbers into whole numbers. To make life easier, let's only look at numbers greater than zero, the numbers 1,2,3,4,. . . . Suppose that we input a number K by marking K consecutive squares, placing the Turing machine's head at the leftmost square, and turning it on. If the machine eventually stops and leaves a string of L consecutive marked squares, then we will say that the machine has taken the number K as input and given the number L as output. The machine T is thus thought of as calculating a numerical function $T(K) = L$. Let's describe the instruction sets for calculating some simple numerical functions.

***T*(*K*) = *K* + 1.** The idea here is to add one mark to the marks on the tape and then stop. There are various sets of Turing machine instructions that will do this. Consider the two-state machine with the instruction set [1*XR*1, 1*BX*0]. The first instruction says, "If you are in state 1, and you are looking at a marked cell, then move one cell to the right and stay in state 1." The second instruction says, "If you are in state 1, and you are looking at a blank cell, then mark the cell and enter state 0." Recall that entering state 0 means the same thing as "turn yourself off." If you start a machine with these instructions at the left end of a string of marks, it will move to the right until it hits a blank. When it gets to the blank, it makes one extra mark and then stops. In general, if the machine is presented with N consecutive marked cells, then it will work for a minute and then stop, leaving $N + 1$ marked cells. We then say that this machine computes the function $T(N) = N + 1$. A faster-running rule for computing the same function is [1*XL*1, 1*BX*0]. A machine with these instructions just adds a mark to the left end of the string, instead of moving all the way out to the right end.

***T*(*K*) = *K* + 3.** Let's describe a five-state machine that adds three marks to the right end of the starting string:

[1*XR*1,	Move to the right past all the marks.
1*BX*2, 2*XR*3,	Make a mark and move one cell to the right.
3*BX*3, 3*XR*4,	Make a second mark and move one cell to the right.
4*BX*0]	Make a third mark and stop.

***T*(*K*) = *K* + *K*.** In the last example we used the states 2, 3, and 4 to "count" the making of three extra marks. In this example, the number of extra marks to be made depends on the particular *K* chosen. The trick is to use the original *K* as a counter and end up with two marks for every mark in *K*. More precisely, we erase the marks in *K* one by one, and for each mark we erase, we put down two. Eight states are needed:

[1*XB*1,	Erase the leftmost mark.
1*BR*2, 2*XR*2,	Move to the blank after the original *N* marks.
2*BR*3,	Skip a space.
3*XR*3,	Move past any extra marks already out there.
3*BX*4, 4*XR*4, 4*BX*5,	When you get to a blank make two marks.
5*XL*5, 5*BL*6	Move left past the spacer blank.
6*BB*0	Stop if the *N* marks are all gone.
6*XL*7, 7*XL*7	Otherwise move left past all the marks.
7*BR*1]	When you get to a blank, move right and start over.

***T*(*K*) is undefined.** It is always possible that a Turing machine may calculate forever without giving any answer. Here we give an example of a Turing machine that goes into a loop and never gives any output at all. Let *T* have two instructions: [1*XL*2, 2*BR*1]. *T* looks at the first mark in the input *K*, moves one square to the left, then moves one square back to the right and starts over. Since *T* never stops, we say that *T* defines no output. In general it is quite hard to predict if a given program might lead to such a loop or not.

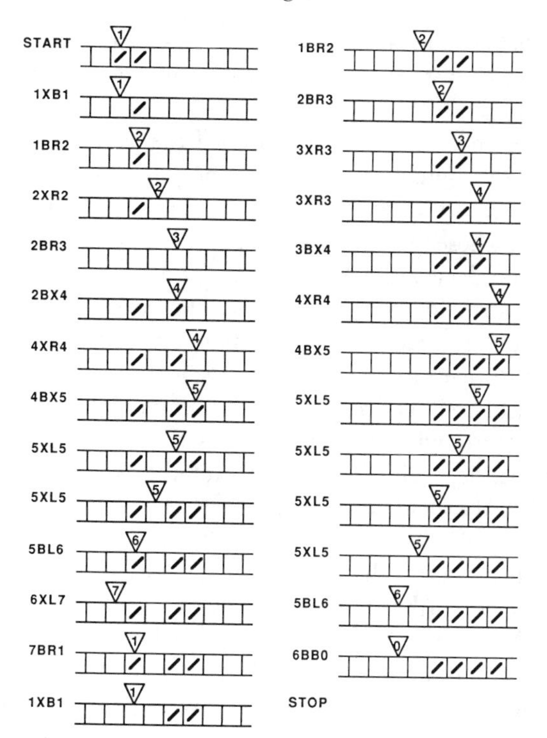

Fig. 116 The $K + K$ Turing machine turns two marks into four marks.

Turing machines are very simple, but it is possible to prove that every digital computer is equivalent to a Turing machine. The essence of a digital computer is simply that it changes individual symbols, one by one, according to certain very simple rules. Digital computation never involves anything more complex than the basic Turing machine activities of (i) reading memory, (ii) writing into memory, (iii) changing the memory location being scanned, and (iv) changing the processor's internal state. In general, if a problem can't be solved by a Turing machine, then it can't be solved by any digital computer at all.

The Turing machines I've described really have three kinds of input: the number of states on the stick that keeps track of which state the machine is in, the program input of instruction quadruples, and the data input of marks on the work tape. We can imagine that all of our machines have the same internal mechanism, consisting of a reader, a look-up device, a tape head, and a stick-mover, which function as follows:

The reader reads the present state off the stick (say it is 3) and scans the present square of the work tape (say it is marked) to come up with the first two symbols of the next applicable instruction (in this case 3*X*). These two symbols are passed on to the look-up device.

The look-up device runs through the list of instruction quadruples to find an instruction that begins with the correct two symbols. If there is no such instruction, or more than one such instruction, then the look-up device moves the state stick to 0, turning the machine off. Otherwise the look-up device finds a unique instruction (say it is 3*XR*4) that starts with the two symbols that were passed on from the reading device. The third symbol of the correct instruction is passed to the tape head, and the fourth symbol is passed to the stick-mover.

The tape head marks the present square, erases the present square, or jerks the tape so that the reader sees the next square to the right or the left, depending on whether the symbol given it by the look-up device is an *X*, a *B*, an *R*, or an *L*. (If the given symbol is *R*, the tape head jerks the tape over one square to show the reader the next square on the right.)

The stick-mover slides the stick along until the square seen by the reader has the same number as the symbol that was passed on by the look-up device. If there is no stick position that shows the correct symbol, then the stick-mover moves the state stick to 0, turning the machine off. (If the symbol is 7, the stick is slid to show the reader the symbol 7).

The generic reader/look-up-device/tape-head/stick-mover Turing machine can be thought of as a universal device that accepts a compound input consisting of total-states/program-quadruples/tape-data. Since this generic machine only has to do a finite number of definite tasks, it will have only some limited number of distinct internal mechanical states.

A machine like this is called a "universal Turing machine." A universal Turing machine *U* takes two tape patterns as input: the program code *P* and the data code *D*. The computation of *U*(*P*,*D*) starts when we give *U* a tape with the marks representing *P*, followed by a blank space, followed by the marks representing D. In computing $U(P,D)$, U simulates the computation of $T(K)$, where T is the machine with program code P.

A universal Turing machine can be built using less than a hundred internal states, and if the program and data are coded up in special ways, the number of states needed for a universal Turing machine can be brought down to ten or less. How can a simple ten-state machine imitate the behavior of machines with thousands or millions of states? The complexity of the target machine's states is replaced by the complexity of the program code P on the tape. Part of the program code is a string of marked cells that is used to simulate the moving state stick: A blank space is moved up and down this string to stand for the current state of the machine being simulated. U is simple in itself; U draws on the complexity of the P to imitate the target machine. If U is not given a complicated program to work on, then U does not do anything complicated.

The processor of a computer is a kind of universal Turing machine, for when we give a computer some software we are really giving it a program. The program is stored on the memory board with the data, and the processor works over the program and data just like a universal Turing machine works over its tape. The processor has a certain simple built-in program, and the universal Turing machine is based on a simple built-in set of universal instruction quadruples. Both are examples of what are known as "universal computers."

What characterizes a universal computer is that, given the proper program input, it can simulate the behavior of any other digital computer. It turns out that *every* digital computer is a universal computer whose powers are limited in practice only by time and space. The time limitation on a computer has to do with how fast it completes each step and how long you are willing to wait for the answer. The space limitation on a computer has to do with how many memory locations it has available. To avoid problems with space limitations, we often assume that a computer is capable of signaling when it needs some extra memory. This is exactly what word processing programs are doing when they display the message "DISK NEARLY FULL!" If we assume that we can always get our computer yet another new blank memory disk, our computer has a memory that is effectively unbounded.

Even though a particular computer's processor is of limited complexity, the computer can carry out arbitrarily complex computations by working a step at a time and keeping track of the steps in its memory. This is exactly analogous to the way that, by keeping track of things on its memory tape, the universal Turing machine U can simulate the behavior of machines that have many more states.

If you don't mind waiting billions of years for your answer, and if you have plenty of extra disks or tapes for memory, any digital computer is as good as any other. For this reason, asking whether a given problem can be solved by a computer is equivalent to asking whether the problem can be solved by a Turing machine.

The somewhat surprising upshot of these considerations is that "being computable" is an absolute, nonrelative concept. If the values of some function $f(K) = N$ can be computed by any digital computer at all, then they can be computed by a universal Turing machine. Given any digital computer with any input, we can always find a program code P and a data code D so that the universal Turing machine's computation $U(P,D)$ leads to the same output. Conversely, if some problem in computation cannot be solved by the universal Turing machine, then it cannot be solved by any digital computer at all.

Unsolvable Problems

Godel's theorem says that whenever M is a consistent, finitely given logical system for mathematics, there is going to be a specific mathematical sentence $G(M)$ that M is unable to prove or disprove. Formalized systems of logic are important because they make the vague notion of "provability" precise. If I know what a system's axioms and rules of inference are, then I can quite mechanically go about listing all of the theorems the system is able to prove.

In practice, people rarely make scientific discoveries by grinding out the consequences of some fixed set of axioms. More commonly, someone will begin to suspect that a given sentence S is true, and will then go about finding the simplest possible assumptions which imply *S*. These assumptions become part of a new theory *M*. The scientific report on such research may then take the form of (1) a concise and elegant description of the theory *M*, (2) an explanation of why the assumptions embodied in *M* are plausible, and (3) a formal proof that the interesting sentence *S* is a logical consequence of *M*.

Godel's theorem says that this process will never come to an end, because no one is ever going to come up with a final and best theory *M* that decides everything. But there is another reason why scientists tend to work backwards, constructing their theories to fit the results they want to prove. The fact is that it is very hard, given a fixed theory *M* and an arbitrary sentence *S*, to decide if *S* is provable from *M* or not. A practical difficulty is that the actual proof of *S*, if there is such a proof at all, may be much too long for anyone to understand. A theoretical difficulty is that we can't predict how long the eventual proof of *S* might turn out to be. It sometimes happens that very simple sentences have very long proofs.

Still, one might hope to find some quick, mechanical way of testing sentences *S* for provability — perhaps one might look for characteristic symbol patterns or the like. The problem of deciding which sentences *S* are provable from a given theory *M* is known as the "Entscheidungsproblem," or the decision problem. In 1936, the logician Alonzo Church showed that the decision problem is unsolvable. This result is known as "Church's theorem." Church's theorem says that there is no easy way to predict in advance if a sentence can be proved from *M*. The best you can do is start proving things from *M* and wait to see if *S* ever turns up. If *S* is in fact *not* provable, you will wait forever, always thinking that a proof for *S* may be just about to show up.

The meaning of Church's theorem becomes clearer if we think of a formal logical system *M* as being a theorem-proving machine. If I am quite precise about describing *M*'s axioms and rules of deduction, then I can actually program a computer to set up all one-step proofs, all two-step proofs, all three-step proofs, and so on. It's a bit as if the axioms are dominos and the rules of deduction describe legitimate ways of setting the dominos down next to each other. A sentence is provable from *M* if there is a domino pattern leading out to it. By working through all possible proof patterns and printing out the conclusions of all the legal proofs, our machine can mechanically generate all of M's theorems. To ask if S is provable from M is the same as asking if the theorem-proving machine M is ever going to print the sentence S.

If it weren't for Church's theorem, we might suppose that there is a Turing machine or computer DM so that whenever S is a sentence, $DM(S) = 1$ if S is provable from M, and $DM(S) = 0$ if S is not provable from M. But Church's theorem says that there is no digital computer that gives correct answers for every case of the decision problem.

If we keep in mind that M can be regarded as a computer, we see that Church's theorem has the form of a somewhat surprising limitation on computers: There is no superprogram that can predict what every computer program will do. Where Godel's theorem tells us that no digital computer M can be programmed to print out all the true sentences about natural numbers, Church's theorem tells us that, given a machine M and a sentence S, there is no way to predict if M will eventually say that S is true.

At almost the same time as Church, Alan Turing proved a closely related result, which is known as Turing's theorem: Given a machine M with a program P, there is no standard way to predict if P will ever finish running. Turing tells us that, as a rule, the only way to find out how long a program is going to run is to sit and watch it, and we may have to sit and watch forever.

Turing's theorem tells us something quite practical. It often happens that a computer takes a while to complete some kind of calculation. So that the user knows that the machine is working, and not broken, the machine often fills up the spare time by making noises, or by displaying a message to say, "WAIT, I'M BUSY." For instance, when an Apple Macintosh is busy on something, it displays a picture of an hourglass. (Newer models use a picture of a clock.) When I first saw an Apple showing such an hourglass, I looked closer to see if the "sand" was moving. I wondered exactly how long I was going to have to wait — ten seconds, ten minutes, half an hour? Could it be that the computation had entered a loop, and that if I didn't turn the machine off my wait was going to last forever? Then I realized that the Apple hourglass is a static icon. It *has* to be static, for Turing's theorem implies that there is, in general, no way for a digital computer to predict in advance how long a given task is going to take it.

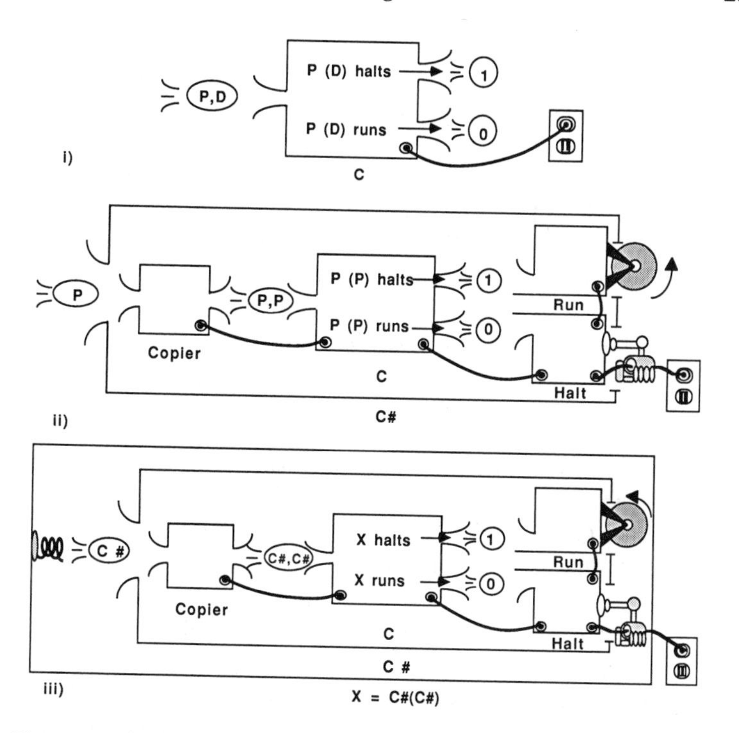

Fig. 117 Turing's proof of the unsolvability of the halting problem. (i) "*P*(*D*) halts" means that the Turing machine with program *P* and data input *D* does a computation and turns itself off. "*P*(*D*) runs" means that the machine *P* with data *D* runs forever. If the halting problem were solvable, there would be a machine *C* that takes (*P*,*D*) as input and outputs a 1 if (*P*,*D*) halts and a 0 if (*P*,*D*) runs. (ii) Given a *C* as in (i), we build a *C#* as indicated. The Copier takes a program *P* and copies it to give a (program, data) pair (*P*,*P*). The Run machine turns on a wheel that runs forever if you give it a 1. The Halt machine has a hand that pulls the plug and turns everything off if you give it a 0. *C#* has the property that *C#*(*P*) runs if *P*(*P*) halts, and *C#*(*P*) halts if *P*(*P*) runs. (iii) Now we build a machine *X* as indicated. *X* performs the computation *C#*(*C#*). If *C* predicts that *X* halts, then *X* runs; and if *C* predicts that *X* runs, then *X* halts. Therefore *C* does not provide a correct solution to the halting problem.

Let's look at Turing's theorem in some detail. Just as Church's theorem says that no computer can solve every case of the decision problem, Turing's theorem says that no computer can solve every case of the halting problem. What is the halting problem? This is the problem of finding a general method to decide whether an arbitrary

Turing machine is going to run forever or print out a result and turn itself off.

A key fact about computers is that not every program works. It is easy to come up with examples of Turing machines that enter some kind of behavior loop and run forever without ever yielding a final result. For a given program P and data set D, we can write $P(D)$ to stand for the computation that the machine with program P performs, given the data D. The halting problem asks if there is a master checking machine C such that, for any program P and any data D, $C(P,D)$ executes a computation with the possible results

$$C(P,D) = 0 \leftrightarrow \text{the computation } P(D) \text{ never halts, and}$$
$$C(P,D) = 1 \leftrightarrow \text{the computation } P(D) \text{ eventually halts.}$$

Turing's theorem states that there is no such program C. His proof works by showing that whenever C is supposed to be a master program-checker C, we can find a computation X such that X runs forever if C predicts that X will halt, and X halts if C predicts that X will run forever. No matter what, C is wrong about X, so this means that C is not a correctly working master program-checker.

Turing's proof hinges on a self-referential trick like the proof of Godel's theorem. Basically, X is a computation that says, "This computation halts if and only if C predicts that this computation will not halt." We get X as follows.

First we use C to build a machine $C\#$ that takes program codes as data inputs, and exchanges the roles of 0 and 1.

$$C(P,P) = 0 \rightarrow C\#(P) \text{ halts,}$$
$$C(P,P) = 1 \rightarrow C\#(P) \text{ never halts.}$$

Then we let X be the computation that takes place if we feed the code $C\#$ to the machine $C\#$. X, in other words, is the computation $C\#(C\#)$.

$$\begin{aligned} C \text{ predicts that } X \text{ never halts} &\leftrightarrow C(C\#,C\#) = 0 \\ &\rightarrow C\#(C\#) \text{ halts} \\ &\leftrightarrow X \text{ halts.} \\ C \text{ predicts that } X \text{ halts} &\leftrightarrow C(C\#,C\#) = 1 \\ &\rightarrow C\#(C\#) \text{ never halts} \\ &\leftrightarrow X \text{ never halts.} \end{aligned}$$

The unsolvability of the halting problem means that there is no computer program that can always correctly predict in advance which programs will work properly. A corollary to this result is that there is no way to estimate how *long* arbitrarily chosen programs will take to run. That is, there is no runtime estimation program R such that, for any program P and any data D, $R(P,D)$ is defined, and $R(P,D) = N$ means that if the computation of $P(K)$ halts at all, then it will halt within N steps.

The reason why there is no such program R is that if we had such an R, then we could use it to build a master program-checker C. Given an arbitrary P and D, we could use R to calculate the number $N = R(P,D,)$, and then we could let the computation $P(D)$ run through N steps. If the computation stopped within N steps we could say $C(P,D) = 1$, and if the computation had not stopped within N steps we would know it would never stop, and we could set $C(P,D) = 0$.

It is this last result that explains why the sand in the Apple "wait" icon doesn't move! More significantly, this result suggests that there is no quick and dirty way to predict an arbitrary program's outcome. Of course, some programs are quite predictable. I know, for instance, that no matter what a given page of text contains, my word-processing program can copy the text in a matter of seconds, but if I have an unknown and somewhat complicated program, then it may very well be that the only way to find out what it will do is to actually *run* it.

This discovery is having some surprising repercussions in the domain of physics. A physical system with very many parts can be thought of as being something like a digital computer with a complex program. Physicists who invent theories are usually trying to find simple ways of predicting what the systems they are interested in will do, but if a physical system is like a universal computer, then Turing's discoveries tell us that there *are* no simple ways of predicting what the system will do!

Many ordinary problems of mathematics can be thought of as particular instances of the halting problem. For hundreds of years now, people have wondered about Goldbach's conjecture. Goldbach's conjecture says that every even number is the sum of two primes. Mathematicians have checked all the even numbers out through one hundred million, and so far each of them can indeed be written as a sum of two primes, but maybe there is some very large even number that is not the sum of two primes? Wondering if there is such a number is the same as wondering if a certain special Turing machine will ever halt. Let G be a machine that goes out through the even numbers one by one, and for each number N checks if there are two prime numbers smaller than N that add up to N. If N is the sum of two primes, then G looks at the next larger even number. If N is not the sum of two primes, then G prints out N and halts. Goldbach's conjecture is true if and only if the Turing machine G never halts. If there were some way to look at G's program and decide whether the machine halts, then there would be a way to decide Goldbach's conjecture.

Each example of the decision problem is also an instance of the halting problem. Suppose I want to know, for instance, if some arcane conjecture can be proved on the basis of the known axioms of mathematics. I design a Turing machine C that examines larger and larger numbers N to see if N is the L27 code of a correct proof of the conjecture. If N does indeed code up a proof, then C prints a 1; otherwise C looks at the next larger N. If I know that C halts, then I know that there is a proof for my conjecture, and if I know that C doesn't halt, then I know that I shouldn't bother trying to prove my conjecture from the axioms at hand.

There are some unsolvable classes of problems that are seemingly much simpler than the halting problem. A good example is the problem known as Hilbert's tenth problem. This problem asks whether an arbitrary algebraic equation has any whole-number solutions. $x^3 - x^2 - 18 = 0$ is an example of an algebraic equation that does have a whole-number solution, 3, and $x^2 - 10 = 0$ is an example of an equation that is not satisfied by any whole number. At one time, Hilbert felt that there should be some mechanical way of looking at an algebraic equation and deciding if it has any whole-number solutions, but fairly recently mathematicians were able to prove that the general problem is unsolvable. That is, there is no program P such that $P(A) = 1$ if and only if the equation A has a whole-number solution.

The Ocean of Truth

Many types of problems can be viewed as problems about deciding which numbers have some special property. Thus, if we imagine coding every sentence up as a number, then the problem of truth turns into the problem of deciding which numbers have the property of being code numbers of true sentences. We say that a property is "computable" if a computer can be programmed to decide whether or not any given number has the property in question. Using the notion of a general Turing machine with program P, we can make this quite precise. A property R of numbers is said to be computable if there is a program P such that

$$N \text{ has property } R \leftrightarrow P(N) = 1, \text{ and}$$
$$N \text{ does not have property } R \leftrightarrow P(N) = 0.$$

In the last section we found that "being the code number of a Turing machine that halts" is not computable. Although the property of being a program that doesn't run forever is not computable, it is "listable." That is, there is a Turing machine that lists, in no particular order, all the programs that will eventually halt. Let me be more precise. A property R of numbers is said to be listable if there is a program P such that

$$N \text{ has property } R \leftrightarrow P(K) = N \text{ for some number } K.$$

I say that the set of all programs that halt is listable because I can imagine setting up a machine that, given any K, tests one program after another, for longer and longer periods of time, until it finds the first K programs that halt and then prints out the code number of the Kth of these programs.

Interestingly enough, the property of being the code number of a true sentence about numbers is neither listable nor computable. Not only can no program be designed to differentiate between true and false sentences about numbers, no program can be designed to list all the true sentences with no false sentences included.

If a property is computable, this means that there is a mechanical program *P* that can look at any *N* and give a yes or no answer as to whether or not *N* has the property. "Being a prime number" is an example of a computable property, for given a number *N*, I can try dividing *N* by each of the numbers less than *N*. If *N* is prime, then none of these numbers will go into *N*, and if *N* is not prime then some number (other than one) will go into *N*. "Being the code number of a tautology" is another example of a computable property. Given the code number of a compound sentence *S*, I can set up a truth table and check if *S* has to be true regardless of the truth or falsity of its various clauses.

So a number property *R* is computable if I can build a Turing machine that reliably detects the presence or absence of property *R*. Metaphorically speaking, the property of being a person carrying metal is a computable one — airport metal detectors can decide, for any given person, whether or not that person is carrying metal. If metal is present, the detector gate lights a red light; if there is no metal, the detector lights a green light.

If a property is listable, this means that there is a mechanical program *P* that can list all the numbers having the property. "Being a code number of a theorem of the theory *M*" is an example of a listable property. Given a theory *M*, I can start systematically deriving all of *M*'s consequences. If the number *N* happens to code a theorem of *M*, then I will eventually get to that theorem, and I will know that *N* has the property of naming a theorem. Some short theorems have very long proofs, so even if I have not yet proved theorem *N*, I always have to wonder if I might prove it later.

Let us imagine a situation in which being "saved" is a listable property. Suppose that, at the end of time, an infinite sea of humanity is standing in front of the pearly gates. St. Peter is a very kind guy, and he never looks at someone and says, "You're damned, go to Hell." Instead, he spiels out an infinite list of possible qualifications for salvation. "OK, if you ever saved someone's life, welcome to heaven. People who gave ten dollars to a bum on the street, come on in. Anyone who was born in 1946 come in, too; I like that year. All right. If you died poor, you can get in. Any skin divers? All skin divers are saved." On and on he goes, through an endless list of possible qualifications for salvation. If you're saved, you eventually find out, but if you're damned, you stand there, uncertain forever.

The distinction between computable properties and listable properties may still seem a bit obscure. Why do I bring such a distinction up at all? The reason is that for most of the formal systems of logic we are interested in, the set of all the systems' theories is listable, but not computable. That is, for any decent logical system *M*, we can list all of *M*'s theorems, but there is no finite mechanical way to look at a sentence *S* and automatically know whether or not *S* is a theorem. That there is no such finite program was proved, as I have mentioned, by Alonzo Church in 1936, and is known as Church's theorem.

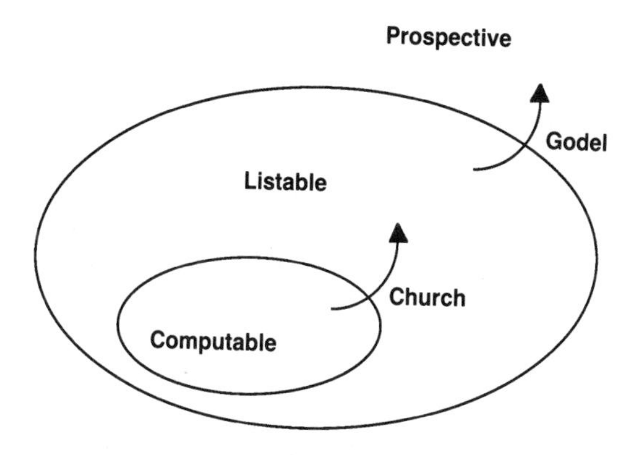

Fig. 118 Church's theorem proves that some listable properties are not computable. Godel's theorem proves that some prospective properties are not constructive.

If we think of all known mathematics as codified in a theory *M*, the property of being an *axiom* of *M* will be a computable property. The axioms we use can be grouped into a finite number of simple schemata; and checking a candidate axiom *S* against the schemata is a mechanical procedure that yields a definite Yes or No in a finite amount of time.

Deciding if a statement is provable from *M* is more difficult, as we have no way of predicting how long a given statement's proof might be. Church's result tells us that, indeed, the property of being a *theorem* of *M* is listable but not computable.

Even more elusive is the property of being a *true sentence* in the mathematical language of *M*. Godel's theorem tells us that this property is not even listable.

Thus we see there are three categories of increasingly complex kinds of property: computable properties, listable properties, and properties that are neither computable nor listable. Suppose that we call a property of the third type "prospective." The set of axioms is computable, the set of theorems is listable, and the set of truths is prospective.

The term prospective is taken from a 1952 paper by John Myhill called "Some Philosophical Implications of Mathematical Logic: Three Classes of Ideas." In this paper Myhill did something very unusual for the uptight formalistic 1950s — he demonstrated that Church's and Godel's theorems can be thought of in a metaphorical way.

Church and Godel tell us that axioms, theorems, and truth lie at three different levels of complexity. Myhill points out that the same three-level split can be found in many other places.

Consider a written page *S*, such as this page, a page covered with letters, spaces, and punctuation marks. Deciding if this page is a passage of grammatical English is a computable matter. One has only to check if the words are all dictionary words, and if the words' parts of speech are such that they can be fit into conventional sentence formats.

Even though the page *S* may be grammatical English, it may very well be meaningless to you. This could be either because the page is gibberish, or because you are unfamiliar with the ideas the page refers to. You are the only real judge of whether *S* is meaningful to you. As a rule, the longer you live, the more different pages *S* will seem meaningful. But it is impossible to predict in advance which pages will eventually become meaningful to you. The property of being meaningful is listable but not computable.

By the same token, the question of whether a page *S* might ever be something you'd want to *write* is a listable one. An author's writings arise, perhaps, in a more or less mechanical way from the endless churning of his or her brain, but there is no way to predict years in advance whether or not the brain in question will eventually output the page *S*. How many times, as a writer, have I thought, "I can't believe I'm saying this!" The same holds true for utterances. That is, the property of a string's being a sentence you might eventually want to say is listable (for all your life you are orally "listing" these sentences), but surely not computable.

Higher properties — such as truth, beauty or virtue — are prospective. There is no fixed rule or token by which we can recognize the true or the beautiful or the good: these human ideals are not computable. Nor is there any kind of program or attitude that will enable any individual person or school to produce all truth or all beauty or all goodness: our highest goals are not to be exhausted by the logical working out of any single system.

Church's theorem is a metaphor for the first fact, the fact that no simple test can give Yes or No answers for important questions. And Godel's theorem is a metaphor for the second fact, the fact that no logical program can hope, even in the limit, to answer all the questions.

After the work of Turing, Church, and Godel, the old dream of capturing all truth in a finite logical net can be seen to be thoroughly bankrupt. Turing's analysis of computation suggests that every finitely given logical system (including human beings) is subject to the theorems of Godel and Church. Godel's theorem tells us that no programmatic method can generate all truth; while Church's theorem tells us that we are unable even to predict the consequences of the programs that we do devise. Is this a cause for despair? Not really. It is rather, I would say, a cause for joy.

A world with no Godel's theorem would be a world where every property is listable — for any kind of human activity, there would be a programmatic description of how to carry it out. In such a world it would be possible to learn a hard and fast formula for "how to be an artist" or "how to be a scientist." It would just be a matter of learning the tricks of the trade.

A world with no Godel's theorem and no Church's theorem would be a world where every property is computable — for any kind of human activity, there would be a fixed code for deciding if the results were good. In such a world an Academy could pass judgment on what was art and what was science. Creativity would be a matter of measuring up to the Academy's rules, and the Salon des Refusés would contain only garbage.

But if there is one thing art history teaches us it is this: all tricks of the trade wear thin, and it's a good idea to keep an eye on the Salon des Refusés.

Our world is endlessly more complicated than any finite program or any finite set of rules. You're free, and you're really alive, and there's no telling what you'll think of next, nor is there any reason you shouldn't kick over the traces and start a new life at any time.

4

INFINITY AND INFORMATION

How Big Is Infinity?

Infinity. The word conjures up tiresome images of endlessness. A dead black space stretching out past the farthest stars. A nightmare library with book after book of unpleasant gibberish. Fever dreams of an endless wall rushing past as one falls and falls and falls. A loop of thought that sends the mind into a fruitless chasing of itself.

But infinity calls forth uplifting images too: God, eternal life, a world without end.

The concept of infinity has always challenged philosophers and mathematicians. Early attempts to develop a precise theory of infinite numbers were hampered by paradoxes. Consider, for instance, two concentric circles, one twice as large as the other. By drawing radii, we can pair up any point on the larger circle with a point on the smaller one, and vice versa. The infinite number of points on the larger circle must be the same as the infinite number of points on the smaller circle, even though the larger circumference is twice as long as the smaller one.

Medieval thinkers like St. Thomas Aquinas concluded from such arguments that infinite numbers are inherently contradictory. Aquinas believed that only God is infinite. Others, like Galileo, suggested there might be infinite numbers, but such numbers would obey rules very different from those that govern finite numbers.

It was not until the late nineteenth century that a comprehensive theory of mathematical infinities was finally developed. This science of infinity, a higher branch of the set theory that children learn in grade school, was invented by Georg Cantor. Cantor spent most of his life as a mathematics professor at a small university in the provincial town of Halle, now in East Germany. Although many of his contemporaries recognized his works as profound and important, he had a number of powerful enemies. Some mathematicians and philosophers questioned the legitimacy of treating infinite sets as definite entities, and a few theologians were concerned that a mathematical analysis of infinity might in some way be blasphemous.

Cantor answered most of these objections in an 1885 essay, which contains the following remarkable passage: "The actual infinite arises in three contexts: first when it is realized in the most complete form, in a fully independent otherworldly being, *in Deo,* where I call it the Absolute Infinite or simply Absolute; second when it occurs in the contingent, created world; third when the mind grasps it *in abstracto* as a mathematical magnitude, number or order type."

For Cantor there were three types of infinity. The first, the Absolute Infinite, is the infinity of theologians. When the average person talks about infinity as inconceivable, he or she is thinking in terms of Cantor's Absolute Infinite. The second sort of infinity concerns the physical universe: an infinite time scale, an infinite number of stars, or matter made up of infinite levels of smaller and smaller particles. The third sort of infinity is studied by mathematicians. The set $\{1, 2, 3, \ldots\}$ of all the ordinary finite numbers is said to be an infinite set of size ω, and the set of all points on the line of real numbers is said to be an infinite set of size c. The symbol ω, pronounced "omega," is used because omega is the last letter of the Greek alpha-

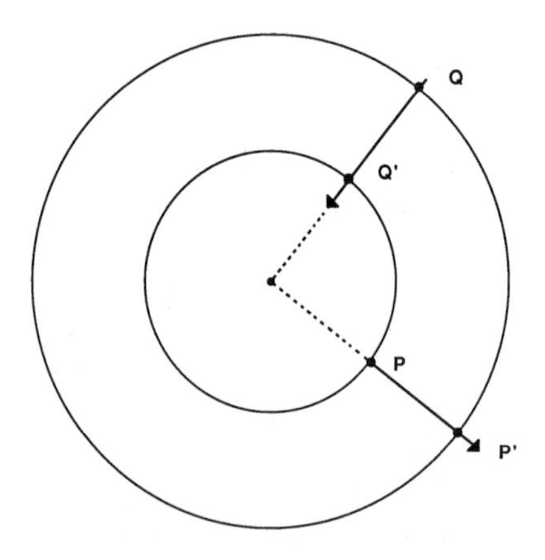

Fig. 119 Radii pair up the points on the two circles.

bet, and the symbol c is used because it is the first letter of "continuum."

Modern set theorists do sometimes try to talk about Cantor's Absolute Infinity as well. They use a capital omega, Ω, to stand for it. The odd thing about Ω is that it is a size level that, by definition, lies beyond any possible means of comprehension; Ω is so big that there is no way for us to single it out. Whenever we think we are talking about Ω, it turns out that we are really talking about some much smaller size level. Put a bit more precisely, if P is some property such that $P[\Omega]$, then there is some A less than Ω such that $P[A]$ as well. This ensures that Ω can't be defined as "the first size level with the property P." This characteristic of Ω is known as the "reflection principle."

A physical analogue of the reflection principle can be seen in the historical fact that every time people have thought they were talking about the whole physical universe, it has later turned out that they were only talking about a small part of the whole. Primitive man's universe turns out to be the planet Earth. The Greeks' universe turns out to be the solar system. The Renaissance universe turns out to be the Milky Way galaxy. Some recent speculations suggest that modern astronomy's universe is a particular space–time bubble, one among many.

Nineteenth-century writers often used the phrase "the Infinite" to refer to God. In terms of vastness and inconceivability, the theological notion of God is indeed close to the mathematical notion of Absolute Infinity. The reflection principle can in fact be transformed into a principle of theology: "God is greater than anything we can conceive." This is really the same as saying, "Whenever we think we have reached God, we have really only reached some level short of God."

The reflection principle, which initially seems wholly negative, has a surprisingly strong positive impact. Consider: Surely Absolute Infinity is bigger than any of the ordinary finite numbers 1, 2, 3, ... Now, we do not want to be able to define Absolute Infinity as "the only number bigger than all the finite numbers," so the reflection principle tells us that there must be a conceivable size level bigger than the finite numbers, yet less than Absolute Infinity. The simplest such level is the number ω, which we mentioned above — a conceivable, garden-variety level of infinity. Moving out past ω, we can tack

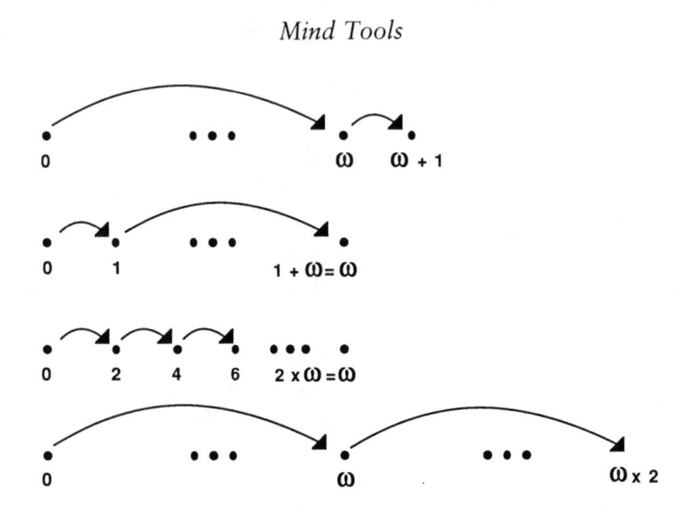

Fig. 120 The operators + and × needn't commute if infinity's involved.

on new levels at will: $\omega + 1$, $\omega + 2$, $\omega + \omega$, and so on. These numbers are known as standard *transfinite* numbers.

The familiar commutative law, which says that $A + B$ is the same as $B + A$, is not true when one is talking about numbers that may be infinite. If there is a possibility that some of the numbers involved may be infinite, we take $A + B$ to mean "First go A steps and then go B steps." We think of positive numbers as being steps to the right and negative numbers as being steps to the left.

The difference between finite and infinite numbers can be stated quite simply. If A is infinite, then A satisfies the equation $1 + A = A$. If K is finite, then $1 + K \neq K$.

Multiplication that involves infinite numbers is also noncommutative. $A \times B$ is taken to mean, "Go A steps, B times in a row." Thus, $\omega + \omega = \omega \times 2$, and $\omega + \omega + \omega = \omega \times 3$.

When might one have occasion to talk about orderings longer than infinity? Are there any kinds of processes that take longer than infinitely many steps? I can think of several examples. Consider the following soap-opera exchange, which has to do with levels of mutual knowledge:

0:	Sue had a secret. Brad came to discuss it with her.
1:	Sue . . . I know your secret.

2:	Yes, Brad. I know that you know.
3:	I know that you know that I know, Sue.
4:	Yes, Brad, I know.
5:	I know your secret. I know that you know I know. And I know that you know that I know that you know that I know.
6:	I know that, Brad.
ω:	Sue, I know, and I know you know I know, and I know you know I know you know I know, . . . this sentence could go on forever, Sue. Through all the finite levels.
$\omega + 1$:	I *know,* Brad. Don't you realize? *I know.*
$\omega + 2$:	I know that you *know,* Sue.
	Oh, Brad. Kiss me, darling. Kiss me hard.
	Kiss you? I didn't know you wanted me to . . . *kiss* you, Sue.

A less frivolous example of the same kind of process comes if a person is meditating and trying to achieve full self-knowledge. It may be that, looking in on yourself, you get into a kind of endless regress of knowing that you know that you know . . . , but then — and this is a distinct intellectual step forward — you find a way to jump out

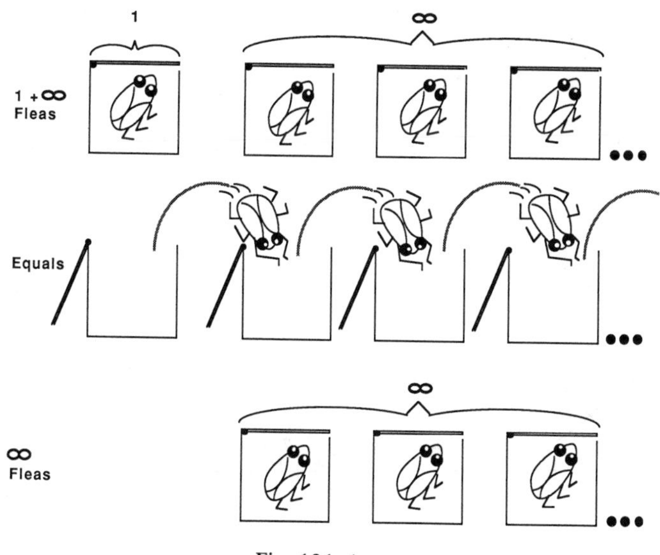

Fig. 121 $1 + \infty = \infty$.

past the whole regress. At this point a kind of ω knowledge is reached, and if you go another step farther, you are at $\omega + 1$.

For a more physical example of a transfinite process, imagine some fiendish game of precision golf in which the ball, the tees, and the holes are mathematical points, and you have a set of clubs consisting of ω irons, one iron for each natural number. The 0 iron drives the ball one meter, the 1 iron drives the ball one-tenth of a meter, the 2 iron sends it one-hundredth of a meter, and in general the *N* iron sends the ball one over ten to the *N*th meters. Suppose that the first

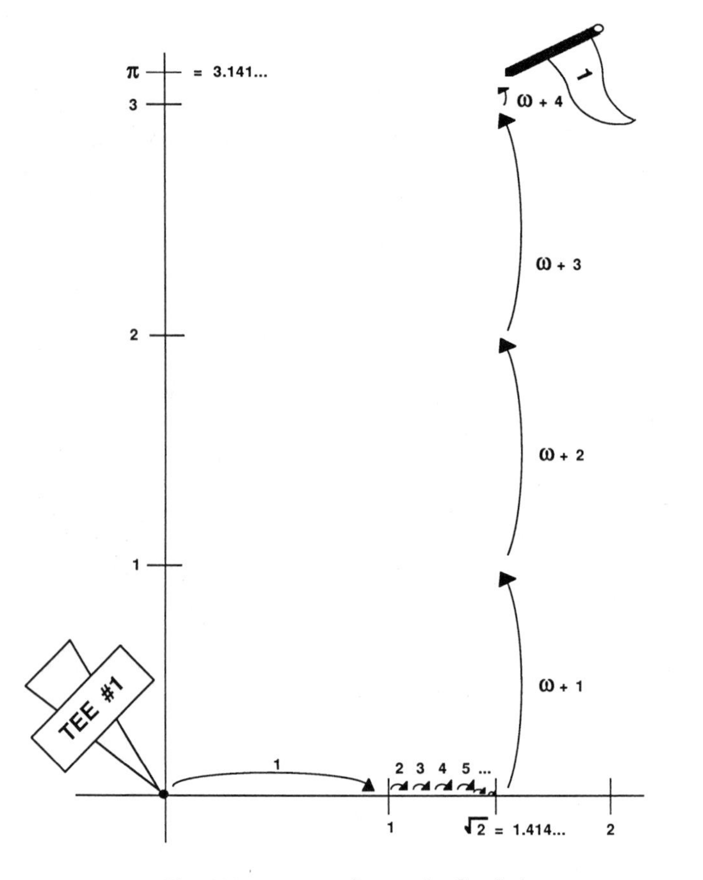

Fig. 122 $\omega + \omega$ strokes on the first hole.

tee is the point (0,0), and suppose that the first hole is at $(\sqrt{2},\pi)$ meters. It will take you at least ω shots to get your ball in, and if you play inefficiently, it could take you ω + ω, or ω × 2 shots! Par for eighteen holes of precision golf would be ω × 18.

As another example of transfinite processes, imagine checking the truth of sentences about the natural numbers. Recall that ∃ and ∀ are called quantifiers, and that $\exists x$ is short for "there exists a natural number x such that," and $\forall y$ is short for "for every natural number y." Any sentence about numbers can be put in the so-called normal form in which there is a string of alternating quantifiers followed by a formula involving only +, ×, =, and names of numbers. Suppose that we are interested in checking the truth or falsity of such sentences by systematically working through all the numbers. I claim that actually checking the truth or falsity of number sentences takes as many steps as shown in this table:

	TRUE	*FALSE*
$(\exists x)P(x)$	N	ω
$(\forall y)(\exists x)Q(x,y)$	ω	ω
$(\exists z)(\forall y)(\exists x)R(x,y,z)$	$\omega \times N$	$\omega \times \omega$

Consider $(\exists x)P(x)$, to start with. We check it by looking, one after the other, at the sentences $P(1)$, $P(2)$, $P(3)$, etc. Each of these sentences takes only a finite time to check. If $(\exists x)P(x)$ is true, then after a finite amount of time N, we will find a number K such that $P(K)$ holds. If $(\exists x)P(x)$ is false, then we will have to run out through all the K's before being sure of this, so it will take us ω steps.

Now look at $(\forall y)(\exists x)Q(x,y)$. If this is true, then for each L, $(\exists x)Q(x,L)$ is true. Checking the truth of each $(\exists x)Q(x,L)$ takes only finitely many steps, so checking the truth of $(\forall y)(\exists x)Q(x,y)$ will involve doing ω finitely long checks, which is ω steps in all. If $(\forall y)(\exists x)Q(x,y)$ is false, then there will be some first L for which $(\exists x)Q(x,L)$ is false. It takes finitely many steps to get to L, and then ω steps to show $(\exists x)Q(x,L)$ false, or ω steps in all.

Now let's do $(\exists z)(\forall y)(\exists x)R(x,y,z)$. For each value M of z that we test, checking $(\forall y)(\exists x)R(x,y,M)$ takes ω steps, regardless of whether it's true or false. If we find a finite value of z that works, then we will know this after $\omega \times N$ steps. If there is no good value of z, it takes $\omega \times \omega$ steps to find out.

This $\omega \times \omega$ is also known as ω^2; ω^3 stands for $\omega^2 \times \omega$, ω^4 stands for $\omega^3 \times \omega$, and so on. We write ω^ω to stand for the first number greater than all the ω^N. A continuation of the argument just sketched above shows that the truth or falsity of any sentence about numbers can be systematically checked in less than ω^ω steps. If it were possible, then, to build a computer that runs for a transfinite amount of time — for ω^ω seconds, in particular — we could build a computer that could decide the truth or falsity of any sentence about the natural numbers. A transfinite computer can partly evade the restrictions imposed by Godel's theorem. That is to say, a transfinite computer can in fact define truth for natural numbers. The basic reason for this is that such a computer's transfinite run time contains a perfect simulation of the natural numbers. If we now want to know the truth or falsity of sentences about the transfinites like ω and $\omega \times \omega$, however, we will need yet more sophisticated machinery. In this sense, our transfinite computer does not evade Godel's theorem. Although it proves all truths about natural numbers, it does not prove *all* mathematical truths.

Let me give one final example of transfinite order patterns. Suppose that we have two functions f and g of natural numbers. For every number x, f and g give us numbers $f(x)$ and $g(x)$. If there is some fixed number K so that for any x greater than K, $f(x)$ is less than $g(x)$, we say that f is "eventually less" than g. If I represent a function as a graph, to say that f is eventually less than g means that the line representing g is, after some point, always above the line representing f. If I represent a function as an equation of the form $y = f(x)$ or $y = g(x)$, then I can arrange some equations into a very long transfinite sequence $y = 0, y = 1, y = 2, \ldots, y = x, y = x + 1, y = x + 2, \ldots, y = 2x, y = 2x + 1, \ldots, y = 3x, \ldots, y = x^2, \ldots, y = x^3, \ldots, y = x^x, \ldots$. The order type of this sequence is $0, 1, 2, 3, \ldots, \omega, \omega + 1, \omega + 2, \ldots, \omega \times 2, \omega \times 2 + 1, \ldots, \omega \times 3, \ldots, \omega^2, \ldots, \omega^3, \ldots, \omega^\omega, \ldots$.

At this point we might naturally wonder if there is some way to make another very large jump in size — a jump as powerful as the jump from the finite numbers to ω? What we need to do is to find some specific way in which Absolute Infinity transcends all the numbers built from ω, and then use the reflection principle to get a new level of infinity that transcends ω in the same way.

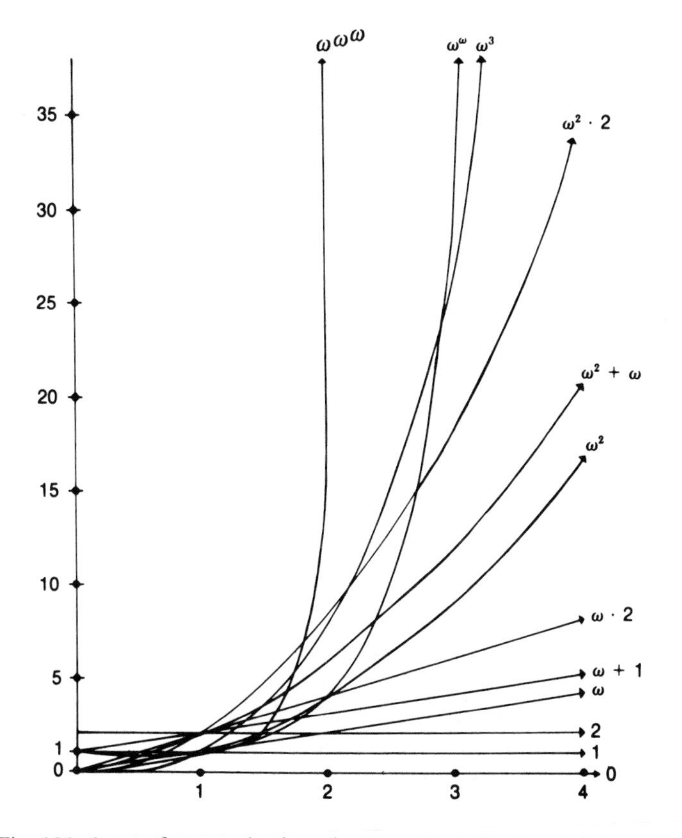

Fig. 123 A transfinite stack of graphs. From Rudy Rucker, *Infinity and the Mind*, Birkhauser, Cambridge, Mass., 1982.

The property we need to talk about is "size." When we want to find the size of a finite set, say a bushel of apples, we count the set up, assigning a number to each apple. The number we reach is the size of the set. If X is an infinite set, we will say that the size of X is the smallest standard number A such that the members of A can be labeled by means of the standard numbers less than A.

When we look at infinite sets, the notion of size has some odd properties. Collections ordered like $\omega + 1$, $\omega \times 2$, or ω^2 members all have size ω. That is to say, I can arrange the finite numbers less than ω so as to fall into the order pattern $\omega + 1$, $\omega \times 2$, ω^2, or even ω^{ω}:

ω order	1,2,3,4,5, . . .	
$\omega + 1$ order	2,3,4,5, . . . ,1	
$\omega \times 2$ order	2,4,6,8, . . . , 1,3,5,7, . . .	
ω^2 order	1, 3, 5, 7, 9, . . . ,	(Odds)
	2, 6, 10, 14, 18, . . . ,	(Multiples of 2, but not 4)
	4, 12, 20, 28, 36, . . . ,	(Multiples of 4, but not 8)
	8, 24, 40, 56, 72, . . . ,	(Multiples of 8, but not 16)

Certainly the inconceivable Absolute Infinity cannot be represented as an arrangement of the finite numbers; to put it differently, Absolute Infinity must have size greater than ω. The reflection principle says that some conceivable level of infinity must reflect this property of Ω. We use the name alef-one, also written as $\aleph_1$, to stand for the first standard infinite number whose size is greater than ω. If we say a number is "countable" if its size is either finite or equal to ω, then we are saying that alef-one is the first uncountable number.

Alef-one is very hard to think about. One way to get at it is to use an analogy, $\aleph_1 : \omega :: 2 : 1$. Getting from omega to alef-one is like getting from 1 to 2. Why do I say this? Is it really so hard to get from 1 to 2?

Yes. Once you get *past* 2, none of the finite numbers is much sweat. If you want to get from 2 to 3, for instance, you simply take two numbers that you already have, namely 1 and 2, and add them to get 3. To form the sum 1 + 2, you don't need any notion of threeness. There are fewer than three numbers being added, and the numbers involved are all less than 3. But if all you have is 0 and 1, you really need the notion of two to get to 2. To form, for instance, 1 + 1 involves knowing how to pick up *two* copies of 1 and add them, but if you don't *have* two yet, then you don't know how to pick up two copies of something.

The analogy to getting from omega to alef-one rests on the fact that if I add together less than alef-one numbers, all of which are less then alef-one, I am adding together countably many countable numbers. It is possible to prove, however, that the sum of countably many countable numbers is always countable. Anything I get by tacking together ω sequences of ω-sized numbers is also ω-sized. Alef-one is the unattainable next size after ω, and 2 is the unattainable next size after 1.

It is also appropriate to compare the jump from ω to alef-one to the jump from finite to infinite. Adding together finitely many finite numbers can never lead to infinity, and adding together countably many countable numbers can never lead to uncountability:

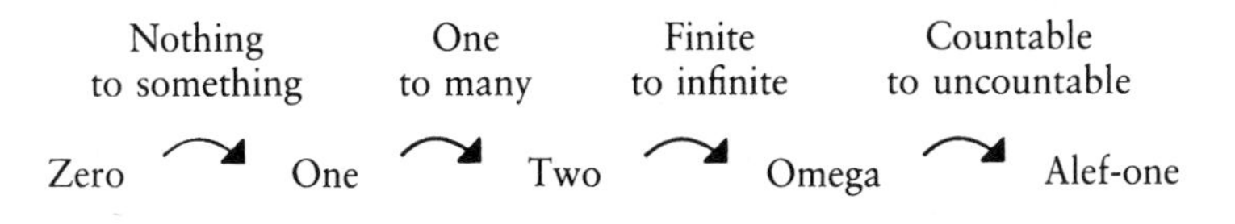

Each of these four jumps is very hard, but looked at in the right way, each of the jumps is inevitable. Each jump follows from applying the reflection principle to Absolute Infinity, which always lies beyond all the steps already taken.

The trouble with our conception of alef-one is that it is very vague. I now want to look at a much more concrete way of getting a set that is uncountably infinite.

Information and the Continuum Problem

In general, if S is a number, finite or infinite, we define 2^S to be the total number of ways in which one can set an array of S switches. Put differently, if we let R be the collection of all possible S sequences of zeros and ones, then 2^S is the size of R.

This 2^S measures the total number of all possible S sequences of zeros and ones. Given that an endless sequence of zeros and ones can be thought of as a binary point fraction, we see that 2^ω measures the size of the set of "endless decimals" naming points on the unit segment [0,1] of the real-number line. For this reason, 2^ω is often called c, as a real line segment is thought of as embodying a *continuum* of infinitely detailed possibilities of location.

In 1873, Georg Cantor proved that c is greater than ω. He showed that there can never be a function that pairs the two sets up in a one-to-one way. Put a bit differently, Cantor shows that if we select one point in the interval [0,1] for each natural number, then there will

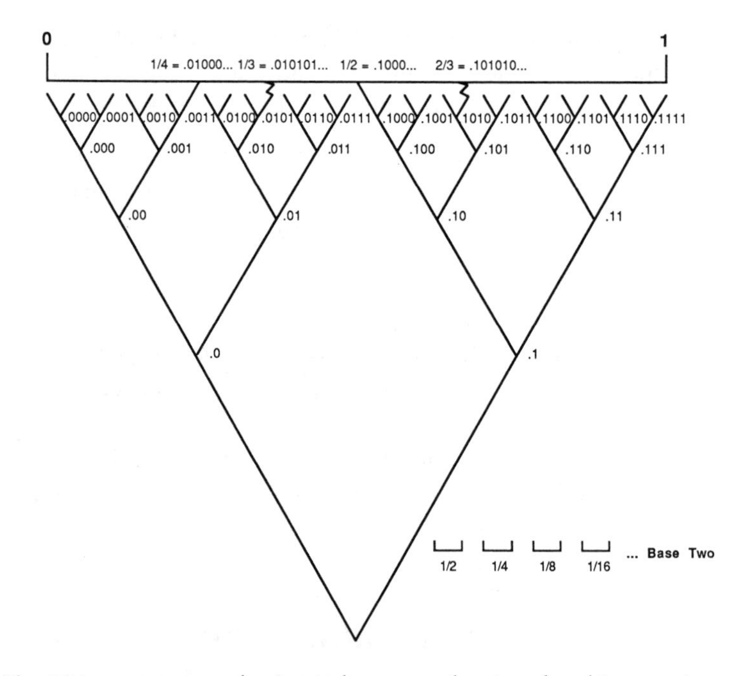

Fig. 124 ω sequences of zeros and ones can be viewed as binary point fractions on the unit interval. From Rudy Rucker, *Infinity and the Mind*, Birkhauser, Cambridge, Mass., 1982.

always be points left over. Given that alef-one is an infinite number bigger than ω, it is natural to ask, Is $c = \aleph_1$? The problem of deciding how *c* relates to alef-one is known as "the continuum problem," which remains unsolved to this day; as a matter of fact, Kurt Godel and Paul Cohen have proved that the continuum problem cannot be solved on the basis of present-day mathematical knowledge.

One difficulty with the continuum problem seems to be that we have so nebulous an idea of what alef-one really is; 2^ω, on the other hand, is quite concrete. The continuum problem can actually be stated without explicitly mentioning alef-one: "Is there a set *X* such that the size of *X* lies between ω and 2^ω?" The idea is that alef-one was supposed to be the first infinite size after ω, so we are asking if *c* is this same first infinite size after ω. Of course it could be that *c* and alef-one are too essentially different to compare, but again, no one really knows. The continuum problem is unsolvable. The difficulty seems to be that the continuum problem does not yet relate to anything that we can really think about. It would be nice if someone could come up with a version of the continuum problem that relates to something as concrete as physics or information theory. That $\log c = \omega$ does suggest that the continuum problem could have something to do with information.

Let's pursue this notion a bit. First let me sketch the method of Cantor's proof that 2^S is always bigger than S. We represent S with the set of numbers less than S, and we represent 2^S by the set of all S sequences of zeros and ones. The idea is to prove that there is no function f that maps the set of S numbers onto the full set of S sequences. The proof works by showing that whenever f is a specific function pairing numbers up with S sequences, there will be a sequence D_f such that D_f is not equal to any of the S sequences hit by f. Suppose we list the sequences that f selects:

$f(1)$ = ① 0 0 0 0 0 0 . . . 0 . . .
$f(2)$ = 1 ⓪ 1 0 1 0 1 . . . 1 . . .
$f(3)$ = 1 0 ① 1 0 1 1 . . . 1 . . .
$f(4)$ = 0 1 0 ⓪ 1 0 0 . . . 0 . . .
$f(5)$ = 0 1 0 1 ⓪ 1 0 . . . 1 . . .
⋮
$f(n)$ = 1 1 1 1 1 0 0 . . . ① . . .
⋮

D_f = 0 1 0 1 1 . . . 0 . . .

The so-called diagonal sequence D_f is formed by going down the diagonal of the square array representing the values of f and changing each zero to one and each one to zero. The point of this is that for any n, D_f is going to differ from $f(n)$ in the nth place. Thus, D_f is *almost* equal to $f(5)$, but the fifth symbol of $f(5)$ is zero, so the fifth digit of D_f is 1. D_f is not equal to any of the $f(n)$, so we know that f does not manage to map the set of S numbers onto all of the set of S sequences. This completes the proof that the size of S is less than the size of 2^S.

This theorem holds for finite S and for infinite S. It proves that two is bigger than one, and it proves that c is bigger than ω. The analogy we talked about earlier can be recast as $c : \omega :: 2 : 1$. I think c is really more natural a concept than is alef-one. The jump from ω to c resembles the jump from 1 to 2 quite closely: c is the number of all possible ω-bit sequences of information, and 2 is the number of all possible 1-bit sequences of information; $2^1 = 2$, and $2^\omega = c$.

The whole essence of twoness comes out of the 0–1 distinction that underlies the modern concept of information. What does Cantor's theorem tell us about information theory?

Suppose that I speak of an ω sequence of zeros and ones as an "infinite bit," or an ω bit for short. I do this in the spirit that I earlier spoke of an "analog bit" as being a location in some kind of space, normally known only to within some dozen bits of precision. An analog bit given to infinite precision is a good example of an ω bit, for the binary tree picture in Fig. 124 shows that an ω bit gives a location to an infinite amount of precision.

It is unknown whether or not our world's various physical features do in fact code up ω bits of information. Certainly some of our world's mathematical features do code up ω bits of information. The set Tr of all true sentences about arithmetic is, as Godel showed, a structure whose information cannot be represented in any finite way. The most obvious way of listing the members of Tr takes ω^ω steps. The set K of the code numbers of all the Turing machines that halt is also infinitely complex, in that no Turing machine can reliably decide which numbers lie in K. K has less information than does Tr, and can be listed by a process taking ω steps.

I think that, if we set aside any old-fashioned preconceptions about the nature of reality and view the world purely as information, it seems likely that each feature of the world does code up, at the least, an ω bit of information. I mean this in the following sense: Given any object, I can ask an endless number of yes–no questions about it. At least it seems as if I could. Now, my interaction is needed to formulate the questions, but it seems that, abstractly, there could be many endless lines of inquiry that exist independently of me. Of course, we don't want to look at question sequences that start repeating the same questions, for then we wouldn't be getting a full infinite bit, but it seems natural, if only as a limiting notion, to suppose that the world is, in each of its aspects, not a game of *twenty* questions, but a game of *infinity* questions.

So let us suppose, just for the sake of discussion, that each object does code up an ω bit of info. Now, how many such bits are there in all? Any two ω bits can be folded up into a single ω bit. That is, if person *A* and person *B* each wish to ask ω questions, then they can take turns, and their answers appear in the single ω bit of the form *AB AB AB AB AB AB*. . . . We can depict this graphically, because there is a binary tree that has one endless branch leading to each point in the unit square.

If there are ω people, it's still no problem. Say the questioners are called *A, B, C, D, E,* and so on. Now all ω of the ω bits can be represented by a single ω bit, by the schema

A AB ABC ABCD ABCDE ABCDEF ABCDEFG ABCDEFGH ABCDEFGHI ABCDEFGHIJ. . . .

If we look at this in reverse order, I am saying that any given ω bit

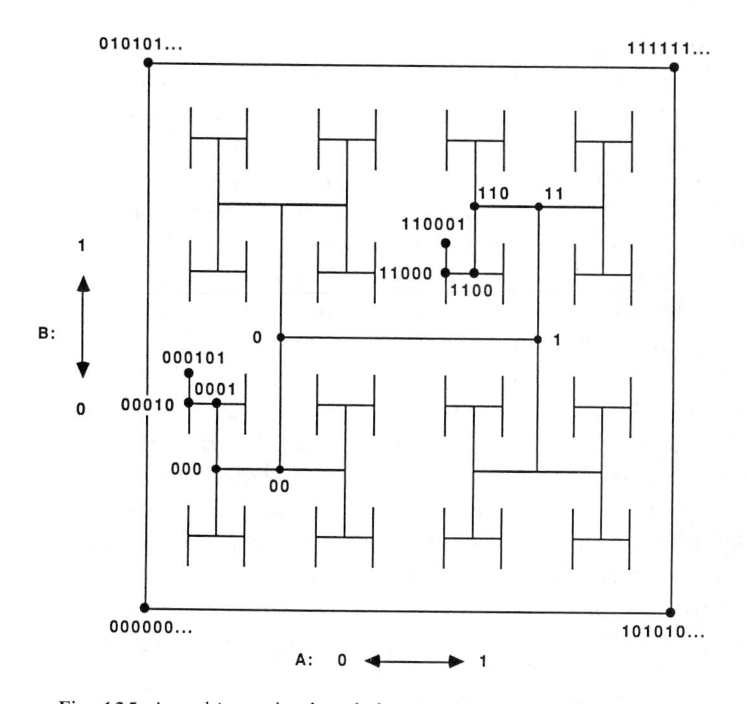

Fig. 125 An ω bit can be decoded as 2 ω bits or a 2-D location.

can be decoded into a collection of ω independent ω bits. For instance, the infinite bit

1 10 000 0101 10111 101111 1100000 01010101 100100101 1100001010 . . .

decodes to

1 1 0 0 1 1 1 0 1 1 . . .
0 0 1 0 0 1 1 0 1 . . .
0 0 1 1 0 0 0 0 . . .
1 1 1 0 1 1 0 . . .
1 1 0 0 0 0 . . .
1 0 1 0 0 . . .
0 0 1 1 . . .
1 0 0 . . .
1 1 . . .
0 . . .

We might get excited now, and speculate that there is some one master ω bit that codes up all the world's information, if only you decode it in the right way, *but this is exactly what Cantor's theorem rules out!* No decoding process, no matter how subtle, can turn one infinite bit into *all* the infinite bits. This is the content of Cantor's theorem. There is no single infinite bit that codes up the cumulative information contained in *all* infinite bits.

The universe has a mocking way of always being just out of one's grasp. Here I have idealized and imagined getting infinite knowledge of something; I have dreamed that I can extract an ω bit of information from the world. Would this make me as smart as God, who knows all things? No, it would not, for if *my* ω bit exists, then, since there's nothing special about me, it seems that all the other possible ω bits must exist as well, and Cantor's theorem shows that the collection of all ω bits contains more information than does any single ω bit. The collection of all minds knows more than any one mind can know.

So, in terms of information theory, the continuum problem is this: *Is there any body of information that is intermediate in size between being (a) so small as to be codable by a single ω bit, and (b) so large as to code up the information in every ω bit?*

It is conceivable that a combined theory of physics and information might have something to say about this question. For instance, suppose that we were to develop a theory under which the fundamental units of reality — the "particles" — are ω bits. In the context of such a theory one might discover a class A of phenomena such that A has too much information to be coded by any ω bit, but has essentially less information than does the whole universe of ω bits. A might be, for instance, associated with a certain class of elementary particles, or with a certain class of idealized observers. In this situation we would have a reason for saying that the answer to the continuum problem as just stated is YES or that, in other words, there are size levels between ω and the continuum, and c is not equal to $\aleph_1$.

Infinitesimals in Perspective

In the first section we spoke of finding a number ω that is bigger than all the finite numbers 1,2,3, ... We then went on past ω to look at $\omega + 1$, $\omega + 2$, $\omega \times 2$, and so on. We did not mention that one can also put new numbers in *below* ω — numbers like $\omega - 1$, $\omega - 2$, $\omega/2$. These numbers, which are infinite but less than ω, are sometimes known as "nonstandard transfinite numbers." They are to the transfinite number world what fractions and irrationals are to the finite number world.

The theory of nonstandard numbers was only developed in the last twenty years. That we can squeeze numbers like $\omega - 1$ in between ω and the finite numbers means that ω cannot really be spoken of as the *first* infinite number. Indeed, there is no such thing. All the numbers greater than the finite numbers 1,2,3, . . . are infinite. Each of them satisfies the equation $1 + K = K$, which characterizes infinite numbers.

One reason people don't like to think about $\omega - 1$, $\omega - 2$, etc., is that there is a feeling that if we have to deal with all these nonstandard numbers, then we'll never make it out to the nice simple infinity ω. In terms of the actual order of thought, however, ω comes before $\omega - 1$. That is to say, although $\omega - 1$ is less than ω, we think of it as being conceptually more complex than ω. We get to $\omega - 1$ by first jumping out to ω and then moving one step to the left.

The important thing to realize is that we do not *count* out to ω in

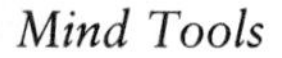

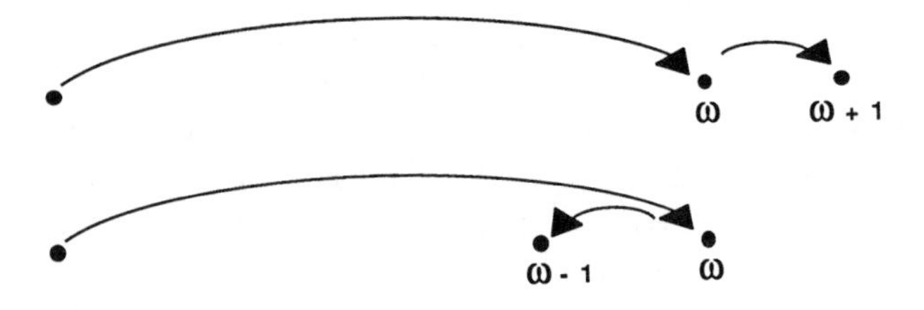

Fig. 126 $\omega - 1$ is conceptually more complex than ω.

any case; we *jump* out there. We use the finite numbers as a ramp to zoom up, and we fly over the nonstandard numbers and land at ω. The view down the road toward Absolute Infinity has fuzzy spots. The numbers we name are hilltops. There's a foggy nonstandard valley between the finite numbers and the numbers near ω. The situation is a bit analogous to the real-number line. We get from, say, 1 to 2 by jumping; if we try to look at each decimal between 1 and 2, we'd never get there. Just because 1.9999 is less than 2 doesn't mean that we think of it before we think of 2.

Not only can we talk about $\omega - 1$, we can also talk about $1/\omega$, which is known as an "infinitesimal" number; $\omega - 1$ and $1/\omega$ are closely related to each other, because $1 - 1/\omega = (\omega - 1)/\omega$. Conceptually, $\omega - 1$ and $1/\omega$ each involves just a bit more information than does ω.

Historically, there has been a great deal of controversy over infinitesimal numbers like $1/\omega$, $1/(\omega + \omega)$, and $1/\aleph_1$, very small numbers not quite equal to zero. Newton and Leibniz talked about infinitesimals in order to invent calculus, but after that infinitesimals fell into wide disfavor. Georg Cantor, who was fighting hard for the acceptance of his own infinitely large numbers, disdained infinitesimals, describing them as "a cholera bacillus that threatens to infect all mathematics."

The eventual proof of the consistency of talk about infinitesimals was carried out by Abraham Robinson in the 1960s. Robinson's proof is based on picturing levels of infinity as arrangements of graphs, as we did in the first section. The idea is that nonstandard numbers like $\omega - 1$, $\omega - 2$, $\omega/2$, and $1/\omega$ can be coherently depicted as functions like $y = x - 1$, $y = x - 2$, $y = x/2$, and $y = 1/x$. Further work on infinitesimals was carried out by John Horton Conway, who in his remarkable book *Numbers and Games* makes a plea for "mathematicians' lib," a viewpoint under which nonstandard numbers and infinitesimals would be treated with the same respect as the standard Cantorian infinities.

Conway bases his argument on an information-theoretic view of numbers: A number is identified with any set of instructions that enables us to compute with it. Instead of worrying if we really *can* squeeze a number like $1/\omega$ in above zero but below 1/2, 1/3, 1/4, 1/5, etc., Conway suggests that we relax and just define $1/\omega$ in terms of the gap between zero and all the $1/n$.

In practice it is certainly true that we can take the reciprocal of any number we can think of. Any upward stretch the mind makes can be mirrored down, and vice versa. There is a very real sense in which 1 is halfway between zero and infinity. If we start at one and keep doubling in size, we slowly approach infinity. If we start at one and keep halving in size, we slowly approach zero. Put a bit differently, the function $f(x) = 1/x$ maps the ray $[1,\Omega)$ onto the interval $(0,1]$.

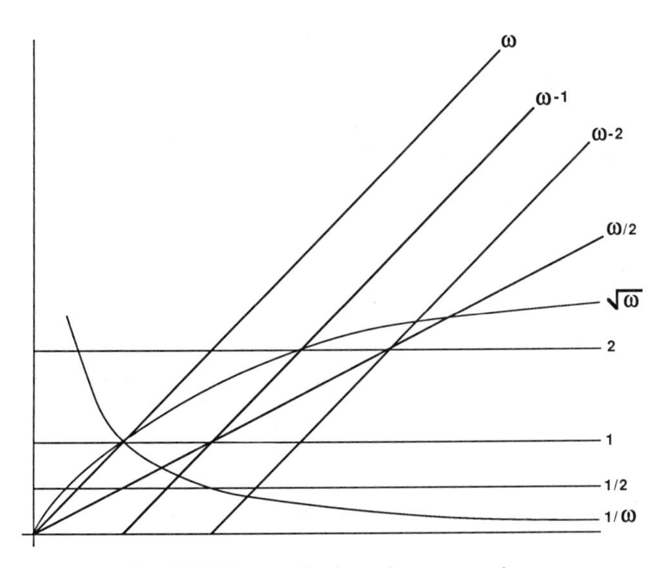

Fig. 127 Nonstandard numbers as graphs.

Taking the reciprocal of a length can be accomplished geometrically by projecting the length x onto a perpendicular unit segment, as shown in Fig. 128. It is interesting here to see that the geometric reciprocal of ∞ is gotten by drawing a line that differs infinitesimally from being parallel to the number line. We'll come back to this in a minute.

Taking reciprocals is easy if we think in terms of exponents. Recall that a 1 with N zeros after it is written as 10^N, and that $1/(10^N)$ is written as 10^{-N}. If we think of numbers as being sizes in meters, and of a person as being roughly one meter in size, we get some interesting

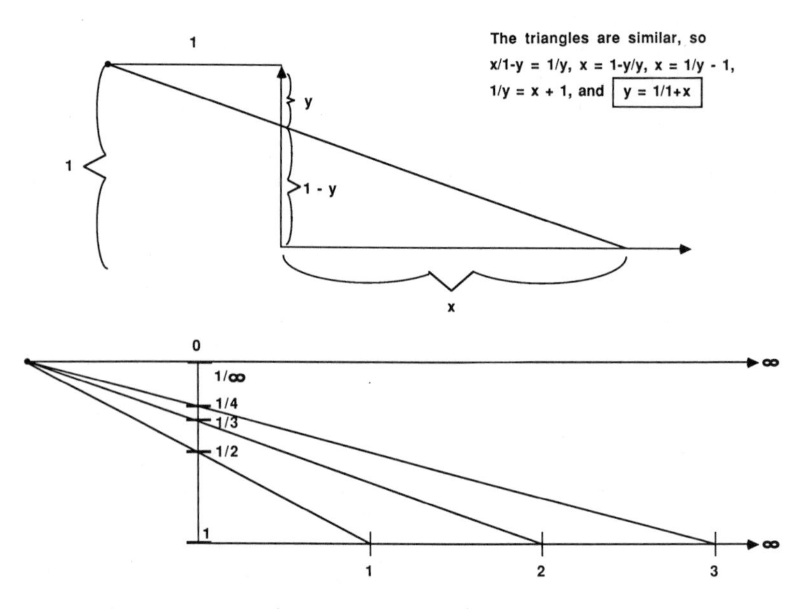

Fig. 128 Geometrical inversion.

ratios. (Although people are closer to 2 meters tall, 1 is easier to work with — and a squatting person is like a sphere 1 meter around.) For instance, the Earth's diameter is about 10^7 meters, and a virus is some 10^{-7} meters across. So compared to a virus, I'm the size of the Earth. Relative to me, the virus is an "infinitesimal" that is the reciprocal of the Earth's "infinite" size. Here are a couple of other approximate reciprocals:

One inch : Five feet	::	Five feet : Hundred yards	
Millimeter : Meter	::	Meter : Kilometer	
Cell : Person	::	Person : Rhode Island	
Virus : Person	::	Person : Earth	
Atom : Person	::	Person : Earth's Orbit Around the Sun	
Proton : Person	::	Person : Distance to Alpha Centauri	

A thumb is to a cheerleader as a cheerleader is to a football field; a cheerleader is the football field's thumb. A nickel's thickness is to a man as a man is to a mile; a man is a mile's nickel. A person is like a cell of a society, but for the whole Earth we're like viruses. An atom is a solar-systemth across, and a proton is a starflightth.

The meaning of this in terms of information is the following. It is just as hard for me to think of a crowd of people big enough to cover Rhode Island as it is for me to think of a crowd of cells big enough to make up a human body. From the viewpoint of 1, one-millionth contains as much information as does one million. Earlier I defined the information in an integer K as being the base-two log of K, which is equal (to within a tolerance of 1) to the number of zeros and ones that it takes to write K out in binary notation. The use of the log function is not appropriate for numbers less than 1, since although a binary point fraction like .100101011100 has sixteen bits of information, the log of this number is something like -1. For numbers less than 1, it's better to forget about logs and just say that the number codes as much information as it takes to write it out as a binary point fraction.

The symmetry between numbers bigger than 1 and numbers less than 1 is especially clear if we imagine that a number is given to us as a finite pattern of marks on a Turing machine tape. A pattern like this will have a form like, say, . . . BBBXBBXXBBXBXBBBBXBXBBB . . . , where there is an endless string of blanks to the left and to the right of the region where the marks occur. Ignoring the infinite blank regions and writing 0 for B and 1 for X, we see that the tape pattern can be represented as a zeros-and-ones string: 10011001010000101. If we want to think of the tape pattern as naming a number, there are two symmetric possibilities: nonzero integers bigger than 1, or nonzero fractions less than 1.

A base-two integer bigger than zero starts with a 1, though it may

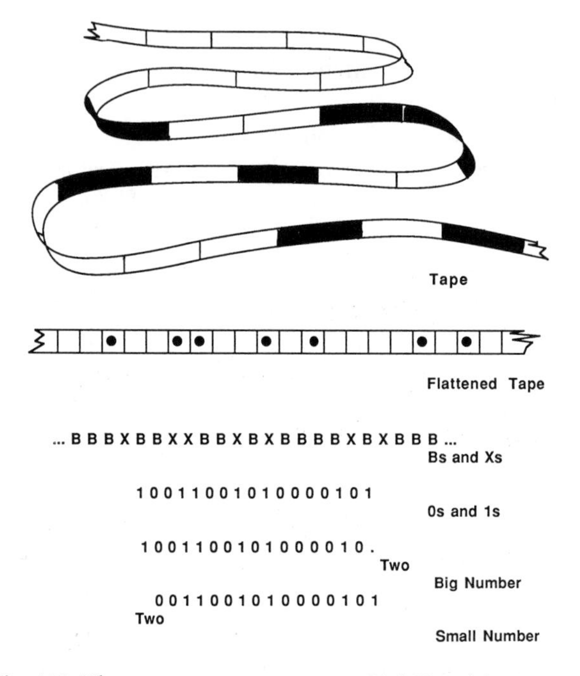

Fig. 129 The same tape pattern names 22,850 and 0.19734192.

have a run of 0s at the end. If we want to interpret the tape pattern as the base-two name of an integer, we replace the last 1 with a point, and we have a binary number like 1001100101000010., which is 22,850 in base ten.

A nonzero point fraction ends with a 1, though it may have a run of 0s at the start. If we want to interpret a tape pattern as the base-two point-fraction name of a number less than 1, we can replace the *first* 1 with a point, and we have the binary point fraction .0011001010000101, which is 0.19734192 in base ten.

The symmetry between increasing and shrinking one's field of attention is brought out very clearly when one sets out to draw perspective pictures. In particular, it is interesting to see how infinitesimals come into play if we try to depict transfinite scenes upon a finite canvas.

Suppose that you want to produce a perspective picture of what

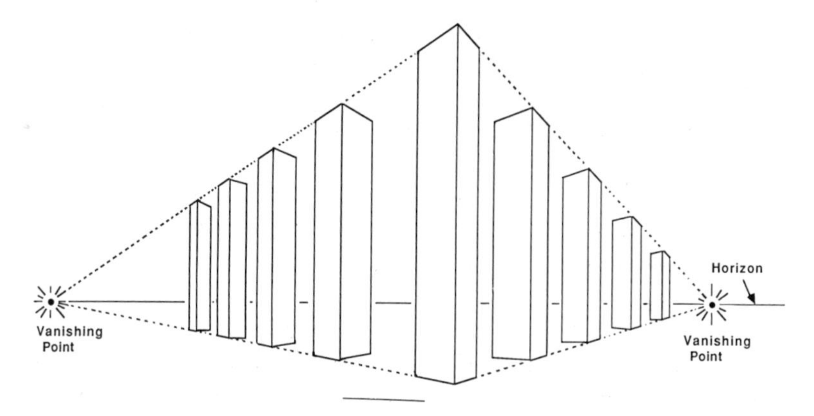

Fig. 130 Perspective scene with two vanishing points.

you see. Extended indefinitely, each line goes out to infinity, though objects may intervene. We think of each line as leading to a separate "vanishing point" on the horizon. If lines are parallel they all go to the same vanishing point. Objects farther away are drawn as closer to their vanishing point, and as smaller. Theoretically the vanishing points are infinitely far away.

The technique of perspective is quite familiar to us now, but it had to be laboriously discovered during the early Renaissance by such Quattrocento Florentine artists as Paolo Uccello and Piero della Francesca. Intellectually, perspective is a breakthrough, because here, for the first time, the physical space we live in is being depicted as if it were an abstract, mathematical space. A less obvious innovation due to perspective is that here, for the first time, people are actually drawing pictures of infinities.

Most of the Renaissance perspective pictures are of buildings, seen head on. Buildings were used because they have lots of parallel lines, which suggest vanishing points. We tend not to think in terms of vanishing points when looking at natural formations such as woods and fields. The reason is that vanishing points are only obviously there in scenes that include families of parallel lines, and such groupings of lines occur only in manmade structures such as roads and buildings. Happening on a house out in the woods is a little like seeing a temple to the human belief in parallelism, with its attendant doctrine of infinitely distant vanishing points. In nature, infinity shows herself as fractals rather than as families of parallel lines.

Although the vanishing point is a definite location on your finite canvas, in reality it is a point that lies an infinite distance away. Ideally, it should lie an Absolute Infinite distance away. If space weren't dirty and warped, any true physical vanishing point would lie Ω miles away. This becomes clear if we think about how a perspective picture is supposed to be constructed. To simplify things, let's have the artist standing on some vast flat desert, rather than floating in space. The ground is thought of as an endless plane, and the artist's eye is thought of as a point some six feet above the ground. A canvas is before the artist. When the artist looks straight out, parallel to the ground, he or she is looking at the "horizon." These horizontal sight lines intersect the canvas along a line that depicts the infinitely distant horizon. Each direction in the plane beneath the artist's feet corresponds to a horizontal line of sight, leading to its own vanishing point. If two

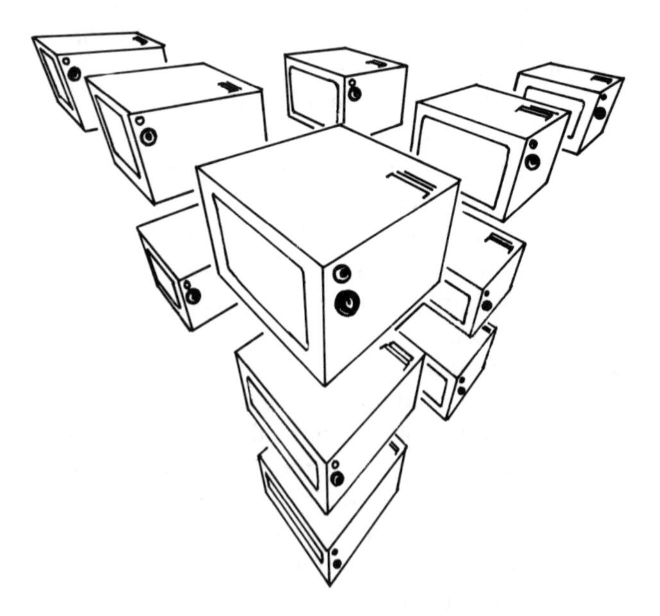

Fig. 131 Three-point perspective in space: ω^3 TVs.

lines run out toward the horizon, the artist draws them as meeting at the vanishing point he or she sees when looking in the same direction as the two lines.

Determining the sizes of objects in a perspective picture is a little tricky. Suppose that, lying on the ground before the artist, there is an endless grid of squares. How to draw them? To simplify things, we assume that one of the grid's lines is perpendicular to the plane of the picture, and that one family of the squares' edges is parallel to the plane of the picture. Uccello realized that the vanishing point of the squares' diagonals should lie on the horizon, just as does the vanishing point of the squares' edges. If we keep drawing diagonal lines from the squares' corners out to the diagonals' vanishing point, we get a perspective picture of the squares. An interesting thing to keep in mind here is that, strictly speaking, none of the lines actually reach their vanishing points. The horizon is a kind of mirage.

What happens if you try to draw a perspective picture looking out at lines that go past ω? Imagine yourself standing in an endless desert stretching out to Absolute Infinity in every direction. Suppose we call it the Ultimate Desert. Standing there, you would see just about what you would see on the bed of some dry salt lake, although things would

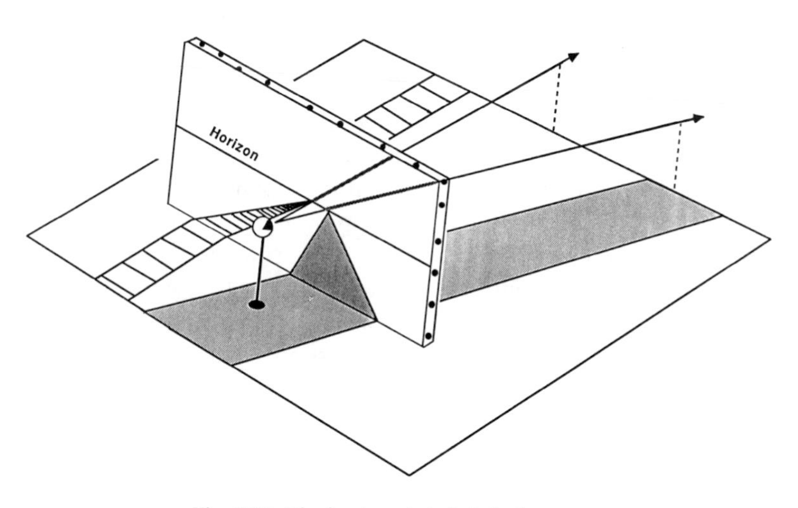

Fig. 132 The horizon is infinitely far away.

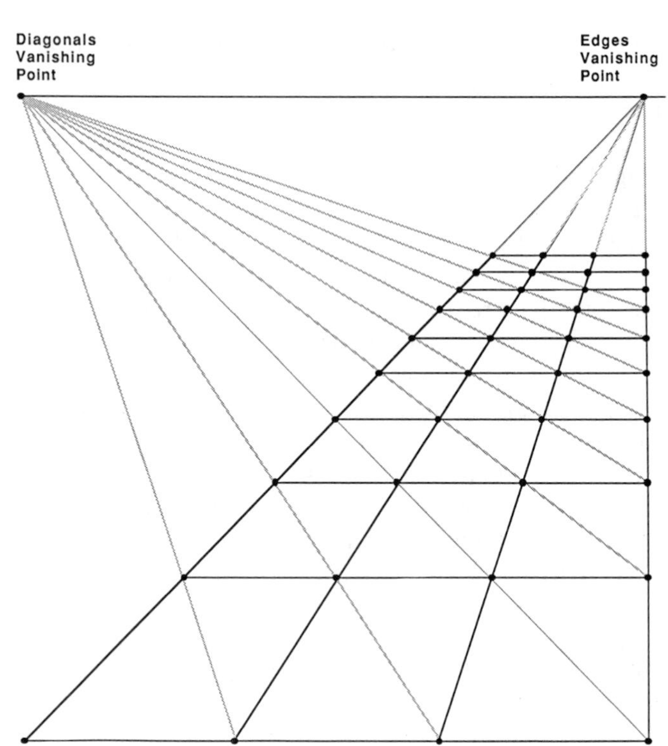

Fig. 133 Equal-sized squares in perspective.

look different right near the horizon. The difference is that the image of any real, earthly desert stops (due to the curvature of the Earth) at a distance of some seventy miles, but the Ultimate Desert runs out past seventy miles, past googol miles, past ω miles, past $\aleph_1$ miles — it runs out past every number short of Ω. Your image of all these extra miles is squeezed into an infinitesimal strip below the horizon.

This is clear if we take a side view of the Ultimate Desert perspector as shown in Fig. 134. For simplicity, suppose that the perspector has unit height and that he stands at unit distance from the canvas. Now, the argument sketched in Fig. 128 implies that the perspector's image of the point at distance X always lies at a distance of $1/(X + 1)$ below the horizon, so the sight line that goes out to ω cuts the canvas at a distance of $1/(\omega + 1)$ below the horizon line. The mark that should stand for ω lies a distance of $1/(\omega + 1)$ below the vanishing point. By the same reasoning, the mark that stands for $\aleph_1$ lies an infinitesimal distance of $1/(\aleph_1 + 1)$ below the vanishing point. Only at the unattainable Ω would the vanishing point ever be reached.

In practice, any picture I can really draw is going to have a minimum pixel size. Here we come back to the concept of bits of precision. The relation between the pixel size and the depictable distances is reciprocal. This is a restatement of our basic insight that any mindstretch up is mirrored by a mindstretch down.

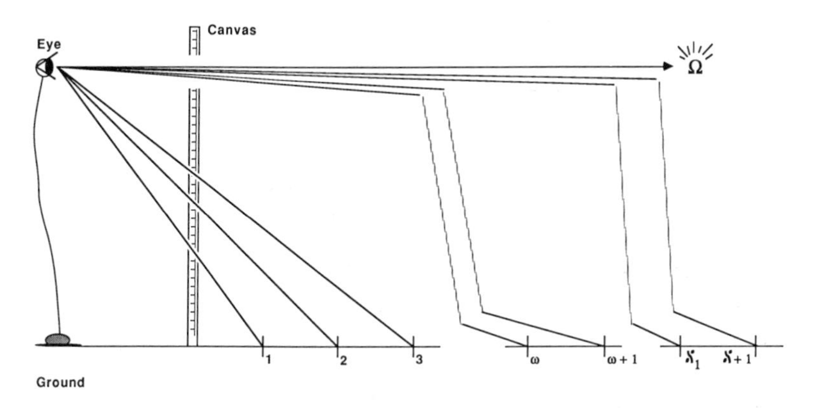

Fig. 134 Transfinite perspective.

Suppose I'm using grains of the sand from the Ultimate Desert as the pixels in the transfinite perspective picture I'm trying to draw. In that case, if I want to get a really good picture, I'd better have sand grains of every infinitesimal size. I'm proposing, in other words, that the Ultimate Desert is nice and fractal, with sand that comes in small, cubelike grains. Let's suppose that the sand grains come in every $1/N$ edge length: 1 inch, $^{1}/_{2}$ inch, $^{1}/_{70}$ inch, one googolth, $1/\omega$ inches, $1/(\omega + \omega)$, etc. I glue these grains to my canvas to depict the lines I want to draw. In order for a line to be fine enough to show the infinitesimal detail required near the horizon, I represent the line as a row of infinitesimal sand grains. It can be shown, for instance, that the sand grain separating the image of ω and $\omega + 1$ should have an infinitesimal size of the order $1/(\omega^2)$.

I kind of hate to leave the Ultimate Desert. Let's take a look at the sand underfoot. We can imagine that the smaller particles are dusted on top of, and in the spaces among, the larger particles. This leads to a very interesting problem, the problem of how many grains of sand there are, not in the whole Absolutely Infinite desert, but just in the finite region directly underfoot.

If we had assumed there were no infinitesimal grains of sand, then we would only need ω of them to fill in a finitely large patch. Once we assume there *are* grains with diameter $1/\omega$, however, we need c of them! This follows from the fact that there are only countably many

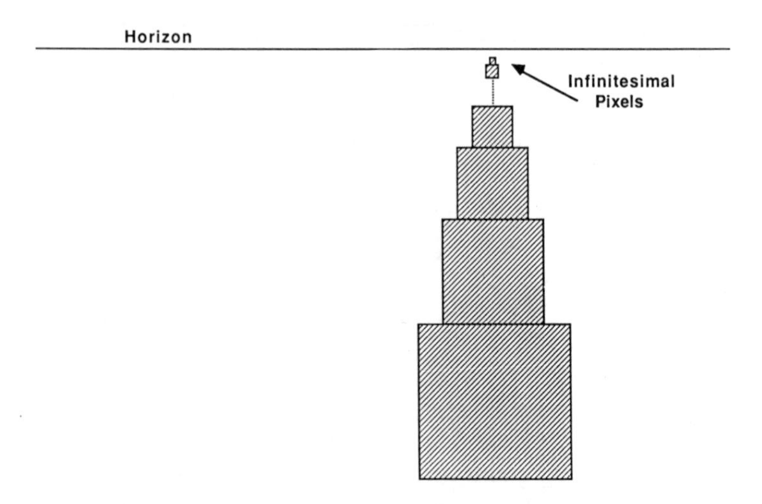

Fig. 135 Using smaller pixels for a finer line.

forks in the binary tree, but there are c, or 2^ω paths up through the tree. The size of the number of grains to be found in any finite region doesn't go up again for a while, because if a is any countably transfinite number, then $2^a = 2^\omega = c$. Going down to grains of size $1/\omega$ changes the number of grains to c, but going on to grains of size $1/(\omega + \omega)$ leaves the number of grains steady at c. If we go on down to grains of size $1/\aleph_1$, then we are going to have room for $2^{\aleph_1}$ of them in any finite region. Is $2^{\aleph_1} = c$? Set theorists don't know, though many of them think that if $\aleph_1$ is indeed less than c, then $2^\omega = 2^{\aleph_1} = c$. This last formula is known as the Luzhin hypothesis. It is a special case of a stronger conjecture known as Martin's axiom. The Ultimate Desert is truly a weird place, and I hope that some of my readers will make their own trips there!

The moral of this section is that there are many similarities between the ways in which we represent information about very large things and about very small things.

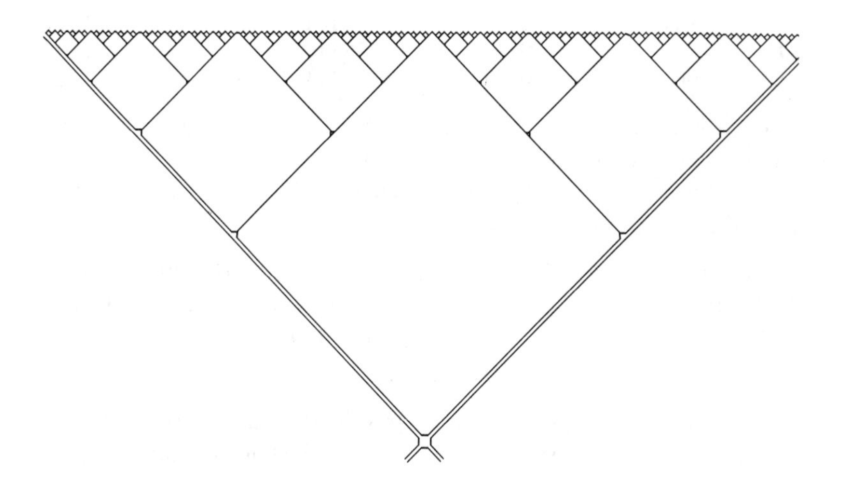

Fig. 136 Fractal sand in the Ultimate Desert.

Algorithmic Complexity

Suppose I find a string of zeros and ones. How much information is the string giving me? This is a tricky question.

If the string is all ones, then I feel it's not telling me much. A random string of N zeros and ones may contain N bits of information, but a repeating string of N ones has no more information than the number N, and recall that in standard numeration systems, the number N takes only $\log N$ digits to write out.

If the string actually codes up an important message, then it will probably look pretty messy. To make this clear, let me show how to code up typed English as zeros and ones. From now on we will call this the "Zeros-and-ones code for English text."

Space	00000	h	01000	p	10000	x	11000
a	00001	i	01001	q	10001	y	11001
b	00010	j	01010	r	10010	z	11010
c	00011	k	01011	s	10011	Capitalize	11011
d	00100	l	01100	t	10100	.	11100
e	00101	m	01101	u	10101	,	11101
f	00110	n	01110	v	10110	'	11110
g	00111	o	01111	w	10111	Return	11111

Each letter is coded by a block of five zeros and ones. A capital letter is coded by preceding it with the Capitalize code, 11011. Thus, the code for "i" is 01001, and the code for "I" is 1101101001. What does the following string say?

11011010010011000000110010111110101000000001100001-
01110000001001000101000010010000000101000100001001-
10011111010000011001011111010111110100100010100000-
100110110100001100101010011100

If you can read this, you're smart. Or at least, you have the code. But look at the string as if you didn't have the code. It looks quite random. How could people ever communicate information if they had to use such a weird code? But, wait. We *do* use a weird code.

Have you ever looked closely at Arabic or Cambodian writing? It looks like a bunch of crazy squiggles. The singer Laurie Anderson is said to have a theory that the Japanese don't really have a language; they just make noises and draw little pictures, which is exactly what *we* do!

Here is the zeros-and-ones code for a very well-known song. To make life a little easier, I'll put spaces between the blocks of five.

11011 01000 00001 10000 10000 11001 00000 00010 01001 10010
10100 01000 00100 00001 11001 00000 10100 01111 00000 11001
01111 10101 11101 11111 11011 01000 00001 10000 10000 11001
00000 00010 01001 10010 10100 01000 00100 00001 11001 00000
10100 01111 00000 11001 01111 10101 11101 11111 11011 01000
00001 10000 10000 11001 00000 00010 01001 10010 10100 01000

00100 00001 11001 11101 00000 00100 00101 00001 10010 00000
10010 00101 00001 00100 00101 10010 11101 11111 11011 01000
00001 10000 10000 11001 00000 00010 01001 10010 10100 01000
00100 00001 11001 00000 10100 01111 00000 11001 01111 10101
11100

If you look closely, you'll see that there's a lot of what information theorists call "redundancy" in this message. Parts of it repeat. Not to keep you in suspense, the message is

Happy birthday to you,
Happy birthday to you,
Happy birthday, dear reader,
Happy birthday to you.

The letter code takes up less space but is basically no more efficient than the code which consists of 505 zeros and ones. Sending letters costs some five bits per symbol, so one hundred letters is no more efficient than five hundred bits, but there *is* a shorter way to send the message:

Happy birthday to you . . . chorus,
chorus,
H. b., dear reader,
chorus.

The shorter English version leads to a shorter code, 345 bits:

11011 01000 00001 10000 10000 11001 00000 00010 01001 10010
10100 01000 00100 00001 11001 00000 10100 01111 00000 11001
01111 10101 11100 11100 11100 00011 01000 01111 10010 10101
10011 11101 11111 00011 01000 01111 10010 10101 10011 11101
11111 11011 01000 11100 00000 00010 11100 11101 00000 00100
00101 00001 10010 00000 10010 00101 00001 00100 00101 10010
11101 11111 00011 01000 01111 10010 10101 10011 11100

Of course, this shorter version will be understood only if the receiver understands the special use of the word "chorus." Suppose we make a somewhat stronger assumption about the receiver's knowledge. Suppose we assume that the receiver realizes that "Happy Birthday" is the name of the familiar song. Then the message can be very compactly English coded as " 'Happy Birthday' for 'reader.' " This has a compact binary code consisting of 160 bits:

11110 11011 01000 00001 10000 10000 11001 00000 11011 00010 01001 10010 10100 01000 00100 00001 11001 11110 00000 00110 01111 10010 00000 11110 10010 00101 00001 00100 00101 10010 11100 11110

If I want to send the message to you still another time, it is enough for me just to say "the song"; this takes forty bits:

10100 01000 00101 00000 10011 01111 01110 00111

By now, if I *still* want to send the message to you, I can just say "it"; that's ten binary symbols: 01001 10100. If all I'm ever going to send you again is the same Happy Birthday message, and you know this, then I could use the most efficient possible code, and send a single bit every time I want to send my message: 1, or for that matter, 0.

So how many bits of information does the Happy Birthday song contain? The simplest measure is the number of zeros and ones it takes to send the message the first way, fully written out. This was, recall, 505 bits. We might call this number the message's "presentation information." Just about anyone can reconstruct the message from the second version, which only takes 345 bits. The 345-bit message can be thought of as an "algorithm," or set of instructions, for generating the full message. The 160-, 40-, 10-, and 1-bit versions are still more compact algorithms that presuppose increasingly large amounts of knowledge on the part of the receiver.

For any given receiver *R*, a message's complexity is the smallest number of bits *R* needs to guess the whole message. More formally, we can call this quantity the message's algorithmic "*R*-complexity." A symbol string's *R*-complexity is the length of the shortest algorithm that enables *R* to generate the string in question.

As we saw above, as long as a person *R* knows the zeros-and-ones code table, the algorithmic *R*-complexity of the Happy Birthday song is no greater than the 505 bits it takes to write out the full binary code of the message, but if *R* knows in advance that the message is going to be the Happy Birthday song, then the message requires only one bit to transmit.

Algorithmic complexity can be made absolute if we fix on some one standard receiver R. The most powerful machines we know of are universal Turing machines. Given enough data, a universal Turing machine can carry out any possible computation. We let U be some one fixed universal Turing machine. Now we can define complexity to mean algorithmic U-complexity:

> **Definition.** If M is a string of symbols, then the *complexity* of M is defined as the length of the symbol string P, where P is the shortest string such that $U(P) = M$.

Recall that a universal Turing machine has the property that, for any digital computer C whatsoever, there is an "emulation code" E such that for any data string D, the computation $U(E,D)$ is equivalent to the computation $C(D)$ that C would perform on D.

Lots of things are universal Turing machines. The first few electronic universal Turing machines people built are known by name: Univac, Joniac, and Maniac were three of them. There's a great photo of the mathematical computer scientist Stanislaw Ulam standing in front of Maniac and looking the part. *Every digital computer is a universal Turing machine,* even the one you bought your kids for $99. The notorious "endless tape" that a universal Turing machine requires has potential existence as the unlimited number of memory cassettes or diskettes that can be bought for the machine. The advantage to having larger computers is not that they can do more, it is that they can do it so much faster, and with so much less tape.

The real significance of universal Turing machines is that it can be convincingly argued that each person is a universal Turing machine. One can even push the argument and argue that trees, atoms, and galaxies are all universal Turing machines. Everything is a universal Turing machine!

But just to fix the ideas you can, if you like, think of our standard universal Turing machine U as being a personal computer with a big hard-disk memory. U has a lot of software already booted up — a good integrated software system that has at its command a word processing program, a spreadsheet program, a LISP program, and a few expert systems. The machine is hooked to a laser ink-jet printer that is always kept loaded with paper, and there is an unlimited supply of extra memory diskettes that the machine can ask you to load into it. Relative to this U, a text's complexity is the smallest number of keypunches that would cause the text to be printed out on paper.

By now the reader may well wonder what I am getting at. Let me explain that the line of thought being described lies in the branch of mathematical computer science known as "algorithmic information theory." Algorithmic information theory was developed in the 1960s by three men, Gregory Chaitin (IBM), A. N. Kolmogorov (USSR), and R. J. Solomonoff (Zator). Inspired by Godel's theorem, Chaitin saw how this line of thought leads to the result now known as "Chaitin's theorem," a very powerful result. One measure of its power is that Godel's theorem can be derived as an easy corollary of Chaitin's theorem. This is impressive if one realizes that when Godel first proved his theorem in 1930, it was widely regarded as the most difficult and profound theorem that any human being had ever demonstrated! And Chaitin's theorem makes Godel's theorem look easy!

OK. So what is Chaitin's theorem, and how do we prove it? Let's go back to our notion of complexity. We have fixed a universal Turing machine U, and we have defined the complexity of any symbol string M to be the length of the shortest program P such that $U(P) = M$. Let's write "complexity(M)" for this quantity. Complexity(M) is the smallest number n such that there is an n-symbol program P that makes U print out M and stop.

Now, what kinds of "messages" M and "programs" P should I look at? What kinds of symbols shall these strings be made of? 0 and 1? 0,1,2, . . . , 9? The whole English alphabet in various fonts? Colored pixels? Whatever "alphabet" of symbols we decide to use, it is important that the set of messages to be generated draws on the same set of symbols as do the allowable programs. U is to be thought of as a kind of amplifier — it takes strings of symbols and makes them more complicated. Thus, depending on which basic alphabet is used, we might think of U as turning short strings of zeros and ones into long strings of zeros and ones, short English texts into long English texts, or simple pictures into more complicated pictures.

In order to have a common standard of comparison, it will be better to limit our symbols to two: a 0 and a 1. If we want to use a bigger alphabet, then we can simply give each symbol of that alphabet a zeros-and-ones code name and write those strings for the letters we want. For instance, if we want to use English, then we can use the code introduced at the beginning of this chapter, and any string of N English letters can be represented by a string of $5N$ zeros and ones.

The presentation information of an N-letter English text is $5N$. We could do the same thing for TV images if we wanted to, assigning code numbers to each pixel and pixel color, etc.

So, now we are thinking of our programs and messages as being strings of zeros and ones. We call these "bit strings." If M is a bit string, then we write "length(M)" to mean the number of symbols in M. Length(M) is what we have already called the "presentation information" of M.

We can assume that our machine U is smart enough that various simple kinds of instructions have simple bit-string codes. In particular, we suppose that there is a particular string that has the force of the instruction "Copy out the following string and stop." If C is this "PRINT:" string, and if M is the message string I'm interested in, I can use the bit string CM as a program P for printing out M. Now C is going to have some fixed finite length c, so I now know that, for any message M, complexity(M) $\leq$ length(M) + c.

The length of a message M provides a first estimate of the information in M, but the complexity of M provides a second, subtler estimate. If M is very repetitive, then the complexity of M will be much less than the length of M. For instance, suppose that M consists of googol repetitions of the string 00001. M might be thought of as the zeros-and-ones code for a string of googol *a*'s. If our machine is reasonably intelligent, it seems that there should be a fairly short bit string that has the force of the instruction: "Print '00001' googol times." In this case we expect that complexity(M) is considerably less than length(M). In general, if a message M is very redundant, or if it says things that the standard universal Turing machine U knows very well, then M will have low complexity. If M is random and unpredictable, then it has high complexity.

Most bit strings are of high complexity. Only one in a thousand binary strings can be generated by an algorithm that is more than ten bits shorter than the string's actual length. Curiously enough, it is very hard to prove that a given string has high complexity.

What we want to do now is to compare the complexity of a logical theory with the complexity of the things it is able to meaningfully talk about. Suppose that T is a reasonably strong logical system, strong enough to prove the familiar facts about mathematics. Suppose also that T is finitely given. This means that T can be thought of as a machine that prints out theorems, the logical consequences of the theory T. We can speak of the whole theory T as having a complexity equal to the complexity of the program of the machine that lists its theorems. This complexity value is simply known as complexity(T). A very simple theory will have low complexity, and an elaborate theory will have high complexity.

We've talked before about the fact that no finite theory can prove everything; that's what Godel's theorem tells us. No consistent, finitely given theory is complete. Now we're almost in a position to improve on this result. It turns out that there's a real sense in which our logic cannot reach out to anything more complicated than what it starts with. Logic can't tell us anything interesting about objects that are much more complex than the axioms we start with.

The following result was first proved in 1965 by Gregory Chaitin.

CHAITIN'S THEOREM

If T is a theory of mathematics and T is

i. finitely given, and
ii. consistent, then
iii. there is a number t such that T cannot prove that any specific bit string has complexity greater than t.

What is the significance of this theorem? To begin with, note that Godel's theorem is a simple corollary of Chaitin's theorem. Since there are endlessly many bit strings, with patterns of arbitrary complexity, the following sentence is certainly true: "There are bit strings whose complexity is greater than t." If we restate this as "There are bit strings M with the property that no t-bit program causes the universal Turing machine U to print M and halt," it is pretty clear that this is really a precise mathematical fact about the kinds of bit strings that exist. But T is unable to prove this mathematical fact, so it must be that M is incomplete. So now, by using Chaitin's theorem, we have established Godel's result, which states that every finitely given, consistent theory of mathematics is incomplete.

The idea behind the proof of Chaitin's theorem is simple, though confusing. Intuitively, the idea is as follows. Suppose I try to reach beyond my abilities by referring to "The first number that I can't name in ten words." The problem is that if the phrase in quotes really means anything to me, then it *is* a ten-word name for the first number that I *can't* name in ten words, and this is a contradictory state of affairs. The proof of Chaitin's theorem is based on a similar kind of paradox.

If T proves the existence of arbitrarily complex bit strings, then "Print the first bit string that can be proved by T to have complexity greater than this program's length" is a legitimate program for generating a specific number. We can choose p to be a big number with a short name (like "one billion"), and then "Write down the first bit string that can be proved by T to have complexity greater than complexity(T) + p" will actually be a name with complexity less than complexity(T) + p. If this program were to describe a message M, then a contradiction would arise, for then M would have a complexity that is both greater than complexity(T) + p and less than complexity(T) + p. To avoid the contradiction, we are forced to conclude that T does not prove the existence of arbitrarily complex bit strings. The t in the statement of Chaitin's theorem can be taken to be complexity(T) + p.

Looked at in one way, Chaitin's theorem says that I can't stand on top of my own head. I can't be smarter than I am. There is a fixed, finite limit to the complexity of the objects about which my logical discourse can say anything significant.

Let me try to make the theorem seem more real. Suppose that the theory T that I use represents the sum total of all the best modern knowledge about mathematics, information theory, and perhaps physics as well, and suppose that the universal Turing machine U that I use has all the best English-language-decoding software there is. Ideally, I'm thinking of U as being about as smart as a person, and of T as coding up all the scientific knowledge that a person might find in the library.

Recall that complexity(T) is defined as being the length of the shortest bit string P with the property that, for any number name N, $U(P,N)$ outputs the Nth theorem of T. Although it seems a bit self-referential, we can also define complexity(U). This would be the length of the shortest program P such that $U(P)$ is the set of Turing machine quadruples describing U's program.

I think it's reasonable to estimate complexity(T) and complexity(U)

as being about one billion each. The estimate of complexity(T) arises from the fact that the zeros-and-ones letter code for a 200,000-word book like *Mind Tools* takes a million bits, and we might imagine that all of modern math and science could be pretty well explained and axiomatized in a thousand such volumes. The estimate of complexity(U) arises from my feeling that a similar-sized collection of volumes would be enough to describe a universal Turing machine U that is able to keep track of, combine, and derive results from the axioms implicit in T. Most of the volumes would simply contain "assembler" programs for converting the other ones into machine language.

Now, the proof of Chaitin's theorem shows that T can't prove that any bit string has complexity greater than complexity(T)$+p$. How big, exactly, is p? It depends not on T, but on the universal Turing machine U on which we base our definition of "complexity." It will, in fact, be equal to complexity(U)$+q$ for some fairly small, fixed q. Given the sophistication of the U that we now have in mind, q would be well under one million.

So, if we think of T as being our best present-day knowledge, and of U as being almost human in its reasoning ability, we get this estimate of the t of Chaitin's theorem: $t \approx$ complexity(T) + complexity(U) + million $<$ three billion.

OK, so what is the theorem telling me about three billion? Suppose that U is confronted with a long and messy bit string M. Suppose that length(M) is something like three billion, and suppose that it has no obvious patterns at all. We might think of M as being a body of data summarizing some observed phenomena of nature. As a scientifically oriented observer, U would like to find a simple "explanation" of M. The best would be to find a "magic formula," a compact thousand-bit string P such that $U(P) = M$. Almost as good would be to find a "book-length" million-bit string P such that $U(P) = M$, or at least a billion-bit "reduction" P such that $U(P) = M$.

After a number of fruitless attempts to find a short "explanation" of M, U turns the problem over to T. The initial hope is that T can prove complexity(M) $= N$ for some N as small as a thousand, or even a million. Even if the proof were nonconstructive, knowing what size explanation of M to look for would be helpful.

T grinds away for a while, but succeeds in proving nothing other than the obvious fact that complexity(M) $\leq$ three billion $+ c$, where, as before, c is the cost of telling U to just copy M out. At this point U begins to hope for a proof that M *has no simple explanation.* U begins to hope that T will somehow perceive the essential randomness of M and prove that complexity(M) = three billion.

Chaitin's theorem tells us that it is *this* that T will never do. If M really is of complexity three billion, then T will never prove that there is some short program explaining M, and T will never prove that there is not any such short program explaining M. T will not tell us much of anything useful about M's complexity.

Now, we may reasonably suppose that the world around us really does contain phenomena that code up bit strings of complexity greater than three billion. Chaitin's theorem tells us that our scientific theories have very little to say about these phenomena. On the one hand, our science cannot find a manageably short "explanation" for a three-billion-bit complex phenomenon. On the other hand, our science cannot definitively prove that such a phenomenon really *doesn't* happen to have a short, magical explanation.

Putting this even more simply, suppose that we think of the whole universe as a phenomenon. Suppose now that M is an immensely long, though finite, bit string that codes up (to a reasonable degree of precision) all the universe that's within a reasonable space–time distance from here and now. Length(M) is very big, perhaps as big as gigaplex. Might there be some short P such that $U(P) = M$?

Throughout history, the occultists and dreamers of science have longed for some simple page of formulas that explains everything. This is the basis for people's fascination with Einstein's $E = mc^2$ formula, and for the present-day fantasy that the new grand unified theories of physics are on the point of "explaining the entire universe." The innocents who think this way are in the position of imagining that the secret of the universe is something to be bought in a ten-dollar paperback, learned in a five-day minicourse, or purchased as a fifty-dollar mantra.

I am by nature unsympathetic to this kind of world view. The notion of a simple answer offends me. I think the world is so full of randomness and unexpected artistic touches that the world code M must have a complexity that is not much shorter than the length of its presentation information. I'm comfortable enough with infinity to believe that the world's complexity is of a very great size.

The killer is that Chaitin's theorem shows that *I can't prove I'm right.* I can't prove that any string at all has complexity greater than three billion, so I can't prove, in particular, that the secret of the universe might not be coded up in some relatively short form.

Strange, strange, strange. The great principle of Plato's dialogues is to follow the argument where it leads. Years ago, I started my study of information theory in the hope of proving the world to be endlessly complex. Yet now, after finally understanding Chaitin's theorem, I find it shows that I can't prove that the universe is more complex than about three billion bits. This isn't what I expected, but it's the truth. The whole value in thinking logically is that it *can* sometimes lead us to unexpected conclusions.

So now I must grudgingly ask: What if there *were* some very simple explanation of the whole universe? How would we account for the great depth that lies between complexity(M) and length(M); between, that is, the small explanation of the world and the big size of the world?

Inconceivability

I say that a pattern is "inconceivable" if it is too complex for me to reproduce in detail. For all practical purposes, the exact arrangement of the stars is infinitely complex, as is the precise pattern of cells that makes up my body. These patterns are not infinite in the strict mathematical sense, but compared to my mental abilities they contain so much detailed information as to lie wholly beyond my grasp. Relative to my information-processing abilities, these patterns are inconceivable.

Suppose I think of myself as being a Turing machine about to make marks on a blank tape. My brain has only finitely many components, and each of these components can be set in only finitely many ways. Imagine all the variable factors in all of my diverse brain components as being controlled by a huge board of, say, three billion on–off switches. By reading, or talking, or absorbing random inputs, I can put my brain in a variety of different overall states. This corresponds to setting those three billion switches in various different arrangements. If each switch can be set independently, that makes for two to the three-billionth, or approximately gigaplex, different overall brain states. Suppose that you start me out in some one brainstate and watch what happens with the passage of time. I may start doing calculations and writing and rewriting things. Just as a Turing machine is allowed to have an endless blank tape to work on, we'll suppose that I have an endless supply of blank paper. I work on and on, setting aside some finished sheets of paper, and eventually I may say, "That's all," and stop adding to what I've already written. I say that some particular pattern of information *P* is "conceivable for me" if there is some initial brainstate that will cause me to output the pattern *P* and then halt. Since there are at most gigaplex initial states in which my brain can start up, there are at most gigaplex patterns that are conceivable for me.

It seems evident that many physically existing structures are inconceivable relative to the human brain. Consider, for instance, trying to learn the specific form and orientation of every blade of grass in a field. Suppose that you had to learn the blade positions so well that you could reproduce them all. You'd be allowed to take as long as you like to get ready, and you could invent all the mnemonic devices you liked. The "getting ready" phase would correspond to setting your brain's three billion switches in the right positions. If the field's blade information were conceivable for you, then there would come a time when you'd be prepared to build an exact replica of the field. Put differently, it would be science-fictionally possible to hook a video unit up to your brain and have the video unit display each square inch of the field in detail. Maybe you *could* master a putting green, but what about the whole state of Nebraska? There's no way.

Plenty of information patterns are inconceivable. Consider the pattern that consists of what everyone on Earth says on one given day, or look up at the sky and imagine being aware of every air current and cloud tendril; imagine being so aware that you can predict how the clouds will look ten minutes from now. A person can only generate a limited amount of information; lots of things are inconceivable.

We have talked repeatedly about a universal Turing machine *U* as being able to perform any possible calculation, given the correct program P. My brain is like a universal Turing machine, but if I am going to work from what is in my brain at any one time, I am not in fact allowed to use arbitrarily long programs. The program for what I am going to do must be stored in the three billion switches of my brain. Roughly speaking, to say that an information pattern X is conceivable by me means that there is a program P such that $U(P) = X$ and P has length less than three billion.

Recall from the last section that we defined a bit string's complexity as the length of the shortest program P such that $U(P) = M$. Given a fixed choice of U, we can define the related concepts of "K-conceivable" and "K-inconceivable":

> **Definition.** A bit string M is K-conceivable if complexity(M) $\leq K$. M is K-inconceivable if complexity(M) $> K$.

Suppose that a human brainstate can be modeled as the combination of a fairly sophisticated universal Turing machine U plus three billion bits of information. In this case, when I say "humanly inconceivable," I really mean "3,000,000,000-inconceivable."

How big, by the way, *is* three billion bits? The newest way of recording sound is by means of the CD, or compact disk. Ordinary recordings code sound in an analog fashion, by means of wavy grooves that jiggle the record player's needle. CDs store sound in digital fashion, by means of tiny yes–no bits that are read with a laser beam. It is said that a half hour of music requires about three billion bits on a CD, so a momentary brainstate involves about as much information as is on a record album. The information on most record albums is highly redundant — the beat is predictable, lyrics are repeated, and so on. A sufficiently sophisticated receiver could probably reproduce the album from a few million of its bits, or less. (Indeed, musical notation is precisely an attempt to describe music with a minimal number of bits — black spots on staves.) But the information in most brainstates is also highly redundant. Often it only takes a few sentences to tell someone exactly what you're thinking. A really complex brainstate is something like a totally unpredictable hour of electronic music.

At this point, I want to talk about inconceivability in a more precise way. Let's assume again that our programs and messages P and M are bit strings of zeros and ones. In order to have an endless supply of possible messages to talk about, let's think of any integer N as being identified with its base-two number name. Thus, twenty is identified with the bit string 10100, and so on. For any possible bit string M, the bit string $1M$ is the base-two name of some number, so the set of all numbers certainly provides us with enough diversity.

Viewed as bit strings, some natural numbers are inconceivably complex patterns of information. The inconceivability of a number has not so much to do with the number's size as it does with the difficulty of naming the number. Gigaplex is very big, but it is certainly easy to conceive of:

$$\begin{aligned} \text{billion} &= 10^9; \\ \text{gigaplex} &= 10^{\text{billion}} = 10^{10^9}; \\ \text{gigaplexplex} &= 10^{\text{gigaplex}} = 10^{10^{\text{billion}}} = 10^{10^{10^9}}. \end{aligned}$$

For me to actually produce the binary code naming gigaplex would be very time-consuming, but there is no information-theoretic difficulty in doing so. I could design a short program and a bookkeeping system so that if I followed the program and kept track with my bookkeeping system, I'd be able (ignoring the likelihood that I'd die or lose interest first) to write gigaplex in binary and halt. As far as information goes, gigaplex only has a few hundred bits. Recall that a single decimal digit requires log 10, or 3.32, zeros and ones, so writing a billion-digit decimal number in binary would result in a bit string 3.32 billion bits long.

If the programs I can use are at most three billion bits long, then it must be that some numbers near gigaplex are actually inconceivable for me. They code up so much information that there is no way I could ever generate them. Although gigaplex itself is conceivable for me, some numbers less than gigaplex are going to be inconceivable for me.

What is the first inconceivable number like? Suppose, that is, that I use W to stand for the first number whose name, in binary notation, is not "3,000,000,000-conceivable." W, in other words, is a number whose binary name I am unable to write out.

At first it looks as if there is a paradox here. If W is the very first inconceivable number, then W is inconceivable and $W - 1$ is conceivable, and you will ask, why can't I conceive of W by first conceiving of $W - 1$ and then adding 1? Given that I have some complex program P that names $W-1$, why can't I form a program $P+1$ that names $(W-1)+1$? Or why can't I get a program Q that names $W/2$ and form a program $2Q$ that names $2(W/2)$?

In both cases the answer is the same. Naming $W-1$ or $W/2$ puts me right up against the limits of my abilities. Learning the program P that generates $W-1$ takes all three billion of my program bits, with no bits to spare. P cannot be made any more concisely than it already is. If I try to tack on an extra little "add one to the output" program, I'm going to have to mess up part of P and I won't be able to get the $W-1$ to which to "add one." By the same token, learning the program Q for $W/2$ uses so much space that there's no room for the "and two times that" program.

The fact is, the conceivable numbers are not closed under the operations of addition, multiplication, exponentiation, and so on. This is a little hard to grasp. The easily conceivable numbers that we normally work with are closed under the usual algebraic operations. If I add, multiply, or exponentiate a few numbers like 3, or 7, or 212, I'll still have a number that I can conceive of. It would be a time-consuming drag to write out 7 to the 212th, but given enough time and enough scratch paper, I could do it. In no way is the calculation beyond my programming abilities. Way out past these small numbers, however, close to W, there is a zone of numbers so hard to conceive of that the sums and products of pairs of two of them may no longer be conceivable.

What lies beyond W? W is inconceivable, but there are some conceivable numbers that are bigger than W. Gigaplex, for instance, has a very short name and, viewed as a pattern of information, is very easy to conceive of, but W, the first (three-billion) inconceivable number, is less than gigaplex. This follows from the fact, which we prove below, that the first K-inconceivable number is less than 2^K, and from

$$2^{3\text{ billion}} = 8^{\text{billion}} < 10^{\text{billion}} = \text{gigaplex}.$$

Not only is gigaplex conceivable, that gigaplex has such low complexity means that gigaplex + N is going to be conceivable for the majority of the N's less than W. Gigaplex $\times$ 2 will also be conceivable, as will $(\text{gigaplex})^2$ and gigaplexplex. Of course, gigaplexplex + gigaplex will be conceivable, too. A diagram of the conceivable numbers is reminiscent of the pattern of the countable transfinite numbers. It

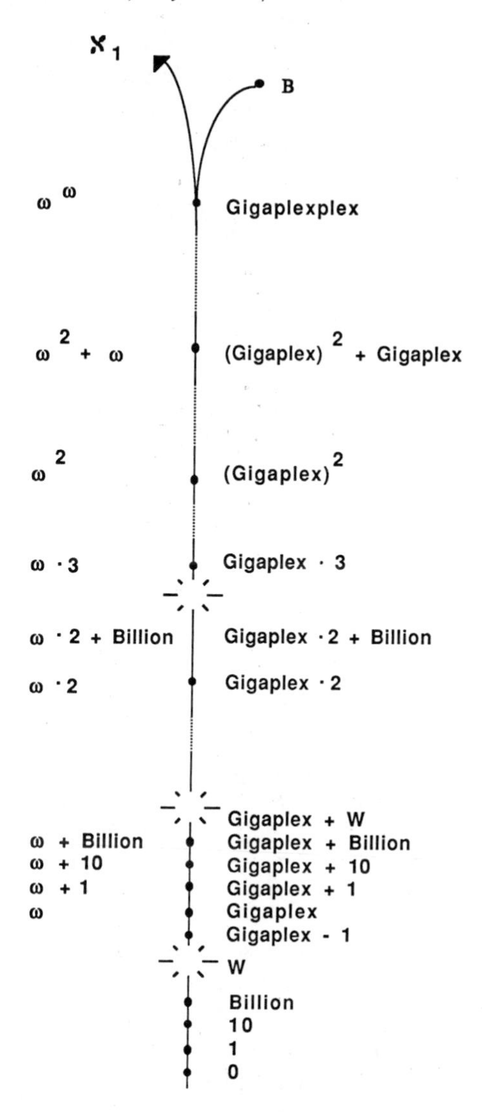

Fig. 137 The conceivable numbers resemble the countable transfinite numbers.

has the same fractal quality of containing runs, runs of runs, runs of runs of runs, and so on. Gigaplex plays somewhat the same role as does ω, with *W* playing the role of the inconceivable "..." gap between finite and infinite.

Since there are at most $2^{3 \text{ billion}}$ (< gigaplex) possible program strings of zeros and ones, there are going to be less than gigaplex conceivable numbers in all. Put a bit differently, if we imagine proceeding ever upward through the numbers, we will eventually reach a point where all possible three-billion-bit programs for generating numbers are exhausted. Suppose that we use B to stand for the very last conceivable number. As it turns out, W and B have very nearly the same algorithmic complexity; each of them lies right on the rim of what we are able to talk about. The fact that B is so much bigger than W means that B might serve as a kind of model of a universe that has a "simple" description.

At this point I'd like to state some definitions and propositions in a precise way. (Unsettled by inconceivability, I reach for the familiar handrail of mathematical formalism.)

Definitions

Let U be a universal Turing machine, and let K be a constant.

Definition 1. If N is a number, complexity(N) is the length of the shortest program P such that $U(P) = N$.
Definition 2. If complexity(N) $\leq K$, we say N is K-conceivable.
Definition 3. If complexity(N) $> K$, we say N is K-inconceivable.
Definition 4. $W(K)$ is the first number that is K-inconceivable.
Definition 5. $B(K)$ is the largest number that is K-conceivable.

An arbitrary K-bit program is simply a list of K zeros and ones, so there are at most 2^K possible K-bit programs P. Any number that is K-conceivable is named by one of these programs, so there are at most 2^K numbers N that are K-conceivable. This means that we have to hit a K-inconceivable number before we get to 2^K, so we know that $W(K) < 2^K$.

Proposition 6. $W(K) < 2^K$.

Let a be a (relatively small) number that represents the cost in bits of telling U to "add one." Thus, if P is a program of length p such that $U(P) = N$, then there is a program P^* of length $p + a$ such that $U(P^*) = N + 1$.

Proposition 7. $K < \text{complexity}(W(K)) \le K + a$.
Proposition 8. $K - a < \text{complexity}(B(K)) \le K$.

Proposition 7 holds because we know $W(K) - 1$ is K-conceivable, while $W(K)$ is not. Since $W(K)$ is not K-conceivable, its complexity is greater than K, but since $W(K) - 1$ is K-conceivable, we know, by our choice of a, that $W(K) - 1 + 1$ is $(K + a)$-conceivable. Proposition 8 holds because we know $B(K)$ is K-conceivable, while $B(K) + 1$ is not. Since $B(K)$ is K-conceivable, its complexity is less than or equal to K, but if $B(K)$ were to have complexity less than $K - a$, then $B(K) + 1$ would be K-conceivable, which is not allowed.

Now let c be a (relatively small) number that represents the cost in bits of telling U to "copy out the following string." Thus if M is a bit string of length m, then there is a program P of length $m + c$ such that $U(P) = M$. We can use this to get an estimate of length$(W(K))$, the length of the bit string that is the binary number name of $W(K)$.

Proposition 9. $K - c < \text{length}(W(K)) \le K$.

I know that the length of $(W(K))$ has to be greater than $K - c$, since otherwise we could use a "copy out" program with length $\le K$ that writes out $W(K)$, contradicting the K-inconceivability of K. I also know that the length of $W(K)$ is less than K, because, by Proposition 6, K is less than 2^K, and any number less than 2^K has a binary name consisting of at most K zeros and ones.

So now, combining Propositions 7 and 9, I know that $W(K)$ has a binary name that is a bit string with length and complexity both approximately equal to K. In other words, the bit string naming $W(K)$ is "random" or "incompressible" in the sense that there's no way of describing it that is much more efficient than just writing the string out.

$W(K)$ is the first string that can't be printed by a K-bit program. Recall that $B(K)$ was the largest number name that *can* be printed by a K-bit program. How does $B(K)$ compare to $W(K)$, as number and as a bit string?

Let e be a (relatively small) number that represents the cost in bits of telling U to "raise to the 2." I want e to have the property that, if P is a program of length p such that $U(P) = N$, then there is a program P^* of length $p + f$ such that $U(P^*) = 2^N$. That is, $U(P^*)$ is the bit string that consists of a one followed by N zeros.

Now we know it takes about K bits of program to write out

Table 4.

1–5	1,	1 + 1,	1 + 1 + 1,	2 + 2,	3 + 2,
6–10	3 × 2,	3 + 2 + 2,	2e3,	3 × 3,	3 × 3 + 1,
11–15	3 × 3 + 2,	3 × 3 + 3,	4 × 3 + 1,	4 × 3 + 2,	4 × 3 + 3,
16–20	4 × 4,	4 × 4 + 1,	3 × 3 × 2,	4 × 4 + 3,	4 × 4 + 4,
21–25	4 + 3 × 3,	5 × 4 + 2,	5 × 4 + 3,	2e3 × 3,	5 × 5,
26–30	5 × 5 + 1,	3e3,	3e3 + 1,	3e3 + 2,	3e3 + 3,
31–35	3e3 + 4,	4 × 4 × 2,	2e5 + 1,	2e5 + 2,	2e5 + 3,
36–40	3 × 2 e2,	2e5 + 5,			5 × 4 × 2,
41–45					5 × 3 × 3,
46–50			4 × 4 × 3,	4 + 3 e2,	5 × 5 × 2,
51–55				3e3 × 2,	
56–60					5 × 4 × 3,
61–65				4 × 4 × 4,	4e3 + 1,
66–70	4e3 + 2,	4e3 + 3,	4e3 + 4,	4e3 + 5,	
71–75					5 × 5 × 3,
76–80					5 × 4 × 4,
81–85	3e4,	3e4 + 1,	3e4 + 2,	3e4 + 3,	3e4 + 4,
86–90	3e4 + 5,				
91–95					
96–100					5 × 5 × 4,
101–105					
106–110			3e3 × 4,		
...					
121–125					5e3,
126–130	5e3 + 1,	5e3 + 2,	4e3 × 2,	5e3 + 4,	5e3 + 5,
131–135					3e3 × 5,
...					
161–165		3e4 × 2,			
...					
191–195		4e3 × 3,			
...					
216–220	3 × 2 e3,				
...					
241–245			3e5,	3e5 + 1,	3e5 + 2,
246–250	3e5 + 3,	3e5 + 4,	3e5 + 5,		
251–255					
256–260	4e4,	4e4 + 1,	4e4 + 2,	4e4 + 3,	4e4 + 4,
261–265	4e4 + 5,				
...					
316–320					4e3 × 5,
321–325				3e4 × 4,	
...					

341–345			4+3 e3,		
. . .					
371–375					5e3×3,
. . .					
401–405					3e4×5,
. . .					
486–490	3e5×2,				
491–495					
496–500					5e3×4,
. . .					
511–515		4e4×2			
. . .					
621–625					5e4,
626–630	5e4+1,	5e4+2,	5e4+3,	5e4+4,	5e4+5,
. . .					
726–730				3e3 e2,	
. . .					
766–770			4e4×3,		
. . .					
996–1000					5+5 e3,
[Lots of numbers skipped in here]					
298,023,273,876,953,125					5e5 e5.

the roughly K bits of $W(K)$'s number name, but it only takes some $\log K$ bits to specify the number K. As $\log K$ is much less than K, $\log K + e$ is also going to be much less than K, so 2^K is K-conceivable. Therefore $B(K) > 2^K$. Combining this with Proposition 6, we get

Proposition 10. $W(K) < B(K)$.

Actually, $B(K)$ is much, much greater than $W(K)$ and 2^K, for I can repeat the instruction "raise to the 2" in such a way that, if P is a program of length P such that $U(P) = N$, there will be programs P^{**}, P^{***}, P^{****}, etc., with respective lengths $p + 2e$, $p + 3e$, $p + 4e$, etc., such that

$$U(P^{**}) = 2^{2^N}, \quad U(P^{***}) = 2^{2^{2^N}}, \quad U(P^{****}) = 2^{2^{2^{2^N}}}, \text{ etc.}$$

Here, $U(P^{**})$ has a binary name consisting of a one followed by 2^N zeros, $U(P^{***})$ is a one followed by $U(P^{**})$ zeros, etc. If we recast the old plex notation in binary terms, then when P names N, P^* names Nplex, P^{**} names Nplexplex, and so on. Given that $\log K$ is much less than K, I can prove lots of results, like:

Proposition 11. $B(K) > 2^{2^{2^K}}$, and length$(B(K)) > 2^{2^K}$.

Remember, though, that the complexity of $B(K)$ is still near K. That $B(K)$ is so long compared to its complexity means that, when seen through the eyes of the universal Turing machine U, $B(K)$ is not at all random-looking. This lack of randomness may be obvious; that is, $B(K)$ may contain lots of long strings of zeros, and certain patterns of zeros and ones may recur in repetitive ways, but it may also be that the simplicity of $B(K)$'s shortest U program is not so easy to spot.

To sum up, for a fixed universal Turing machine U working with K bits of program input, the smallest inconceivable bit string W is a random string of about K zeros and ones, and the longest conceivable bit string B is a much longer string that may look superficially random, but is really the result of a very long computation acting on a short program.

To get a feel for the kinds of patterns conceivable numbers come in, let's look at a kind of toy model of this section's set-up. To make it more readable, lets say that U's programs and messages are written not in some zeros-and-ones "machine code," but rather in an alphabet that uses the symbols 1, 2, 3, 4, 5, +, ×, and e, where I understand AeB to mean A exponentiated to the B power. When two operation symbols are present, we assume that the operations are carried out from left to right.

Now let's look at the numbers that can be named with five or fewer symbols. In other words, we're thinking about "five-conceivability." If we always try to use the smallest number symbols possible, we end up with a set of number names like Table 4.

The numbers 1–37, 40, 45, 48–50, 54, 60, 64–69, 75, 80–86, and 100 are the numbers up through 100 that are "five-conceivable" in this setting. Continuing up through 1000, we get some more, as indicated. The largest number name here is 5e5e5, which is 298,023,273,876,953,125, or some 298 quadrillion.

Names such as 5 × 4 × 3, which use the maximum five symbols, do not lead to any further names, but short names like 4e3 lead to runs of additional names. As was mentioned above, it is generally true that the pattern of numbers conceivable by a given scheme takes on a kind of fractal form. There will be groups of runs separated by long gaps. The groups of runs are made of subgroups of runs broken by subgaps, and so on.

In our five-symbol setting, $W(5) = 38$ and $B(5) = 298,023,273,$-

876,953,125. The complexity of 38 seems to be 7 (with a seven-symbol program, 2e5 + 5 + 1), and the complexity of 298,023,-273,876,953,125 is 5. $B(5)$ is surprisingly complicated to look at. That $B(5)$ is so much more elaborate than its definition has to do with the fact that actually computing the digits of $B(5)$ from its program takes a long time.

The question of how long a program takes to run is fraught with complications and calls for a section of its own.

Runtime

The "runtime" of the computation $U(P) = M$ is the number of discrete steps that U takes to carry out the computation. How exactly do we define the number of steps in a computation?

If U is a Turing machine, we say that U carries out one computational step each time it makes a mark, erases a mark, or moves its tape one square. The runtime of $U(P) = M$ is the number of these steps that the computation requires. Going back to Fig. 115, in the "Turing Machines" section, we can see that the $K + K$ Turing machine T takes 27 steps to compute $T(XX) = XXXX$.

If you offer to pay the owners of a large mainframe computer U to carry out the computation $U(P) = M$, they will bill you on the basis of the computer time the computation takes. A good-sized computer carries out something like a million operations per second, so if one is billed for three seconds of computer time, this means that the calculation $C(P) = M$ has a runtime of about three million. One used to hear horror stories of people who wrote what they thought was a fast-running program, only to find themselves billed for hours or even days of computer time (though now that computer time is cheap, folks hack fearlessly). As we proved in the section on unsolvable problems, it is very hard to predict how long a given program is going to run or if, for that matter, it is ever going to stop running at all.

It is easy to imagine modifying our universal Turing machine U

into a machine U^* that has an extra timekeeping tape on which it makes a mark each time a computational step is taken. If $U(P) = M$, then $U^*(P) = (M,N)$, where N is the number of marks on the timekeeping tape.

Definition. Let U be a fixed universal Turing machine, and let P be a bit string. If there is an M such that $U(P) = M$, we set runtime(P) equal to the number N such that $U^*(P) = (M,N)$. If the computation of $U(P)$ never halts, we say runtime(P) is undefined.

In the last chapter we proved that the halting problem is unsolvable. That is, we proved that there is no program that can correctly predict whether or not a given computation will ever come to an end. As a corollary, we proved that there is no program R that can correctly predict an upper bound on a program's runtime. In terms of our new definition, this result can be stated as follows:

There is no Turing machine T such that, for every program P, T(P) = runtime (P). Put a bit differently, there is no runtime-prediction program T such that, for every P, $U(T,P) = \text{runtime}(P)$. This means that the "runtime" function is what is known as a noncomputable function.

Closely related to the runtime function is the so-called busy beaver function. Suppose we fix some value of K and look at all the possible K-bit programs. We feed these programs to U and watch how long they run. Some programs won't run at all, some take just a little while to run, some take a long time, and some run forever. We define beaver(K) to be the longest finite runtime of any K-bit program. The program that runs the longest is sometimes called the "K-bit busy beaver," because it is so active.

Definition. Beaver (K) is defined as the largest value of runtime(P) achieved by any K-bit program.

Like runtime, beaver is not a computable function. Various researchers have tried calculating beaver(K) for various K and various basic universal Turing machines U, but no one has gotten very far. The function grows unbelievably rapidly. Beaver is closely related to the function B that I talked about in the last section; given all the values of beaver, I can compute all the values of B and vice versa. This implies that the B function must itself be noncomputable. Since $W(K)$ is provably of complexity K, Chaitin's theorem implies that W is also noncomputable, while we're at it.

Perhaps the most interesting use of the runtime notion is in the IBM scientist Charles Bennett's definition of "logical depth." Bennett defines a bit string M's logical depth as the number of computational steps U needs to compute M from M's shortest program. A string with a short "explanation" that requires a great deal of work to understand is logically deep. A string with only obvious, easy to decode explanations is shallow.

> **Definition.** Let M be a bit string, and let P be the shortest bit string such that $U(P) = M$. Length(M) = (the number of symbols in M), as before, and complexity(M) = length(P), as before, and depth(M) = runtime(P).

We have already seen that it is not always easy to estimate how many computational steps it really does take to carry out a calculation of the form $U(P) = M$, but we do know that figuring out and writing down the bit string M will take *at least* length(M) steps. Writing out a K-bit string has to take at least K steps, one step for each bit. Therefore depth(M) $\geq$ length(M).

In general, we say that M is shallow when depth(M) is close to length(M) in size. As we will see below, the depth of M can range as high as virtually any computable function of the actual length K of M — as high as B (K/2), to be precise. It might be, for instance, that M can be named by the program P: "Compute the binary expansion of pi out to the googolth place, and print the last billion digits of this expansion." The digits of pi are so random that, in all likelihood M has no shorter program than P, but computing M from P takes more than googol steps, so the depth of M is greater than googol. This is very much greater than the length of M, which is only a billion.

Bennett, who is a chemist turned information theorist, argues that it may be appropriate to characterize living organisms as physical structures that code up as bit strings with depths much larger than their lengths. Bennett's motivation for this conjecture is the fact that a fertilized egg's genetic DNA is a kind of "program" that generates a complex organism by a process not unlike calculation. Computer-simulating the way in which a human fetus grows from its DNA is an inconceivably difficult problem, runtimewise. Length(baby), the presentation information of a newborn child, is certainly very high, but depth(baby), the number of individual steps involved in growing a baby from its DNA, is quite a bit larger.

"Oh, just *look* at Tom's baby! Isn't it *deep?*," is what I thought when I visited Tom Toffoli's family and held their newborn babe.

Bennett and Chaitin, who are colleagues, speak of the two extremes of complexity as the *crystal* and the *gas*. The atoms of a crystal have the property that they are very obviously arranged according to a simple rule. They are like soldiers on a parade ground. The atoms of a gas have the property that they are totally disordered, and are not arranged according to any rule much shorter than an actual listing of the atoms' positions. Patterns that we find interesting — things such as living organisms and manmade artifacts — lie midway between the extremes of crystal and gas. One of the reasons Bennett invented the notion of logical depth is that he wanted depth(gas) and depth(crystal) to be small, but depth(organism) to be high.

Let's see how well the depth measure realizes this desideratum. Suppose we fix some fairly large value of K and compare the depths of all the various possible K-bit strings. In this context, a crystal is like a string X consisting of K ones in a row, and a gas is like a totally random string G; $W(K)$ is an example of such a string G. The complexity of X is close to log K, and the complexity of G is near K, but they both have minimal depth. The crystalline X and the gassy G both have depth near K. X has a minimal program of the form "Print K X's." This straightforward program is about log K bits long and takes about K steps to run. G has a minimal program of the form "Print the bit string G." This equally straightforward program is about K bits long, and also takes about K steps to run. In terms of complexity, the crystal and gas are quite different, but in terms of depth they are the same. Both are as shallow as a K-bit string can be.

If, as in the last section, W is the first K-inconceivable number, then length(W) and complexity(W) are both approximately equal to K. This means that W is a "gassy" random string whose depth is also about K. If B is the largest K-conceivable number, then, as we know, B is indeed very large, and its length is quite large as well. This means that B has depth much greater than K. Is B a good example of a logically deep object?

Actually, no. There are other bit strings with the same long length as B that are much, much deeper. Compared to B's huge length, the K-bit program for generating B is tiny, and the runtime of this tiny program isn't all that much bigger than length(B).

Bennett has a weird argument for proving that some K-bit strings are very deep. How deep? Just about as deep as you like. How about Kplexplex steps deep? I choose Kplexplex, or $10^{(10^K)}$, simply because it is a humongous number with low complexity. Any other computable function of K will do. The hardest thing about writing a program for Kplexplex is expressing K, which takes complexity(K) bits [$\leq \log K$]. Can Bennett prove that some K-bit string is Kplexplex steps deep? Yes, provided that K is at least several times as large as complexity(U). Bennett told me a program P that says get a simulation of U and do the following: (1) Write out all possible bit strings S of length $K-1$ or less. (2) For each of these S, run the computation $U(S)$ for Kplexplex steps and take what's on the tape by then as an output. (3) Now find the first K-bit string that doesn't appear as an output in stage 2.

If we run the program P, it generates a deep K-bit string called D. We know it generates something, because there are more K-bit strings than there are "program" strings of length $K-1$ or less. This guarantees that there will indeed be K-bit strings that don't appear as outputs in stage 2. The length of P is something like complexity(U) + complexity(K). Given our assumption that complexity(U) is well under K, this means that the length of P is less than K and also, in particular, less than $K-1$. P is a bit string such that $U(P)$ eventually generates D, but stage 2 of the program ensures that P won't generate D in less than Kplexplex steps.

Complexity(D) is necessarily less than or equal to the length of the program P that generates D. If we are working with a high-complexity U and with an easily namable K of low complexity, then length(P) is about the same as complexity(U), so

$$\begin{aligned} \text{length}(D) &= K; \\ \text{complexity}(D) &\leq \text{complexity}(U) < K; \\ \text{depth}(D) &> K\text{plexplex}. \end{aligned}$$

To get the full import of Bennett's result, let's interpret it in the same way we interpreted Chaitin's theorem. Think of K as being the three billion bits we've been treating as characteristic of a person's mental capacity, and think of U as being a fairly simple computer, with a complexity of only a million bits. Bennett's result shows that some strings M have length three billion, complexity under one million, and a logical depth of gigaplexplex.

This means that there are some three billion bit strings M that can be generated by relatively short programs P, but the runtime of the computation $U(P) = M$ is greater than gigaplexplex. If we imagine that M might code up some individual person's total knowledge, then Bennett's result says that it may happen that the knowledge M can be theoretically generated from some simple set of rules P, but it takes an impossibly long time (gigaplex steps) to get from P to M. "I understand you, man, but you'll never know why," says P to M.

We might instead take K to be the million or so bits it would take to write out the most important equations of physics, and let U be a quite simple Turing machine with a complexity of one hundred thousand. Bennett says that it could happen that all the present-day equations of physics can be derived from a tenth as many equations, but that the derivation would take an impossibly long time. There *are* easy answers, but we're too dumb to understand them.

Speaking more loosely, Chaitin showed that we can't prove that the world has no simple explanation. Bennett showed that the world may indeed have a simple explanation, but that the world may be so logically deep that it takes an impossibly long time to turn the explanation into actual predictions about phenomena.

To make it even simpler: Chaitin shows that we can't disprove the existence of a simple Secret of Life, but Bennett shows that, even if someone tells you the Secret of Life, turning it into usable knowledge may prove incredibly hard. The Secret of Life may not be worth knowing.

Everything Is Information

Let's take a look at the physical world around us.

Speaking quite superficially, we can say that any physical object is a system that processes information. Given stimulus P, the system U gives the response M. Throw a rock in a pond and you get ripples, but no one normally thinks of a pond as being like a computer.

One big difference between the two is that a computer acts in a serial, step-at-a-time fashion, but a system like a pond seems to do many things at once. The ripples spread out from the rock in every direction at once. If the pond can be thought of as carrying out computations, we should say that it is computing "in parallel," calculating many things at once.

Although it has been very useful in the past, a Turing machine may not be the best model for real-world computation. The machine is serial, but the world is parallel. Information engineers are presently trying to design parallel computers, and there is hope that these machines may lead to some incredible breakthroughs.

Fortunately, we do possess a very good set of models for parallel computations: the cellular automata. Back in chapter 2, I talked about the game of Life, which is an example of a two-dimensional cellular automaton, or "a 2-D CA" for short. Recall that in Life one starts with a 2-D grid of cells, some of which are shaded black. As time goes by, the pattern of shaded cells evolves in surprising ways. The rules of Life describe only one possible 2-D CA. One might think of other possible rules, with different conditions for killing off a cell, or where a rule might decide a cell's future state not by looking just at its nearest neighbor in each direction, but by looking at the cell's *two* nearest neighbors in each direction.

The key aspect of a CA is that the *same rule is used at each cell site*. The particular CA rule is usually a very short program *P*. The grid's appearance after *N* units of time is determined by the starting pattern, by the CA rule *P*, and by the time *N*.

The game of Life is a 2-D CA that only allows two possible colors per cell: black and white. One can also talk about CA's that allow a whole range of colors — eight of them, or 256. We speak of a CA as being "*K*-state" if *K* colors per cell are allowed. If you look at the color flea circus on a TV screen that's not tuned to a channel, you can imagine that you are watching the evolution of a four-state 2-D CA — four states because each color TV pixel is either black, red, green, or blue.

In fact, a TV's flea circus isn't the output of any simple CA, as it includes signals based on things like static and cosmic rays. It is said that every hundredth speck of flea circus is a signal caused by a photon left over from the Big Bang! One *can* get beautiful color TV pictures of four-state 2-D CA's in action if one buys a color monitor, an IBM PC and a special "Cellular Automaton Machine" or "CAM," which

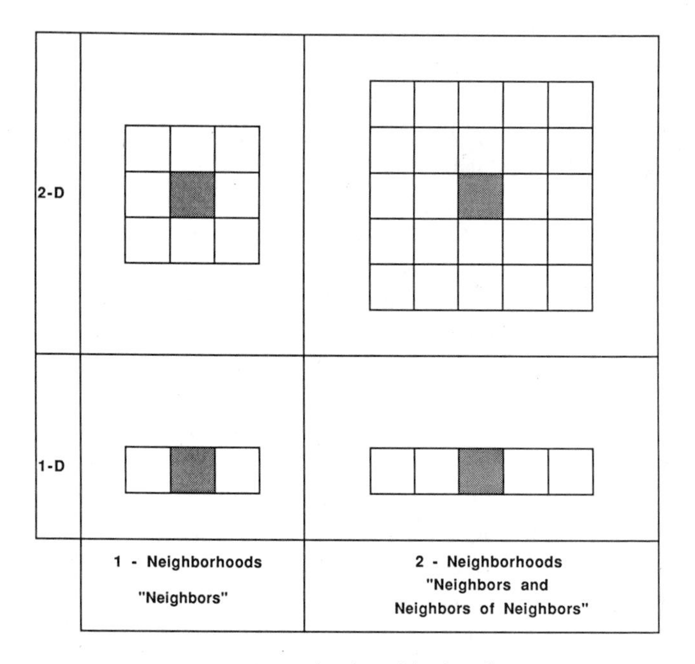

Fig. 138 Kinds of neighborhoods.

is a board of circuitry that plugs into the computer. The CAM is the invention of Tommaso Toffoli and Norman Margolus of the MIT Laboratory of Computer Science, and retails for about $2,000.

Last spring I visited Toffoli and Margolus, and they showed me their CAM in action. It was a stunning experience. Margolus and Toffoli had programmed a variety of interesting four-state 2-D CA rules into their machine. One of them, called "the square rock," starts as a large square of red cells in the middle of the blank screen. The square expands out to the edges of the screen and bounces back. As the bouncing continues, the patterns interfere and form intricate op-art patterns, enough patterns in thirty seconds to make a respectable one-man show.

The monitor used by Toffoli and Margolus had 256 × 256 pixels, or 65,536 cells in all. To avoid the problem of not having an infinite plane of cells, the CAM pretends that the screen's left side touches its right side, and that its upper edge touches its lower edge — just as in arcade video games like Asteroids.

A typical CAM program is thirty-two bits long, and the screen usually gets pretty interesting within a thousand steps and very messy within a million steps. Imagine a situation where we start with a block of nine colored pixels in the screen's upper left corner. Specifying the colors takes two bits per pixel, or eighteen bits. Suppose that, after a million steps, the screen is totally filled with a pattern of colored specks called S. Let's estimate the length, complexity, and depth of S.

$$\begin{aligned}\text{length}(S) &= (2 \text{ color bits per pixel}) \times (64{,}000 \text{ pixels}) \\ &= 128{,}000;\end{aligned}$$

$$\begin{aligned}\text{complexity}(S) &= \text{complexity}(X) + \text{complexity}(P) + \text{complex-} \\ &\quad\ \text{ity}(\text{million}) \\ &= 9 + 32 + 20 \\ &= 61;\end{aligned}$$

$$\text{depth}(S) = (\text{million steps}) \times (64{,}000 \text{ pixels}) = 64 \text{ billion.}$$

On my trip to see Toffoli, Margolus, and Bennett, I also visited Stephen Wolfram at the Institute for Advanced Study, in Princeton. Wolfram, who was a Cal Tech professor at 20 and a winner of the MacArthur prize at 21, has made extensive investigations of one-dimensional CA's. By now he works somewhere else.

In a one-dimensional, two-state CA, we start with a row of cells, some of which are shaded black. As time goes by, the pattern of shaded cells changes in various ways. Generally, CA patterns have *low* complexity but *high* depth. They result from a simple rule plus lots of computation. The row of cells is essentially the same as a Turing machine tape, and we can in fact regard a Turing machine as a multistate 1-D CA. Let me explain how.

Imagine shrinking a Turing machine to the size of a little red triangle with a number in it to stand for the machine's state. Now imagine a Turing tape with some marks on it, and imagine setting the little red triangle down on one of the marks. As time goes by, the triangle darts this way and that, erasing some squares and filling some in.

This is a bit like a 1-D cellular automaton, but not quite. In a CA, we don't want to have anything actually *moving around.* One way to avoid this would be to have a little red Turing machine triangle

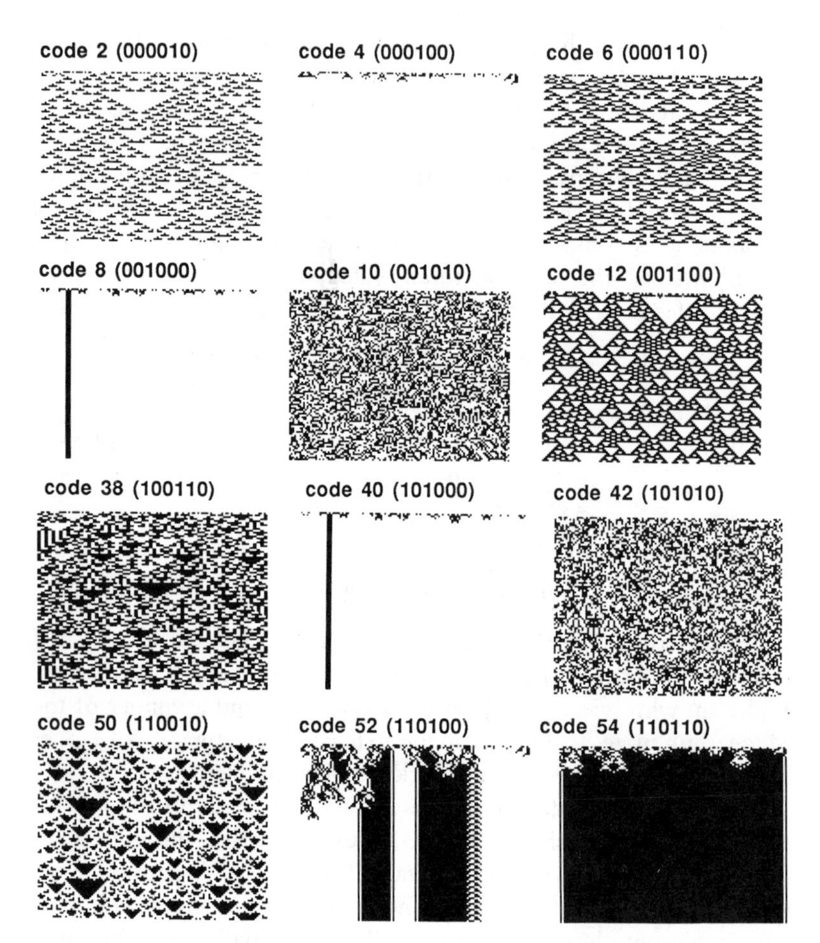

Fig. 139 What various 1-D CA's do with the same random input as they work down the page. Black squares are ones and white squares are zeros. Summing up the values at each site and at its two nearest neighbors on either side gives a number between zero and five. These "legal and totalistic" CA's determine the next value at each site by a rule based on a simply coded look-up table. Picture from Stephen Wolfram, "Universality and Complexity in Cellular Automata," *Physica,* 10D, 1984.

inside each cell, and suppose that each red triangle is wired to its two neighbor triangles. At any given time during a computation, all the triangles except one would display state zero. They would all be "off." The single "on" triangle would correspond to the place where the

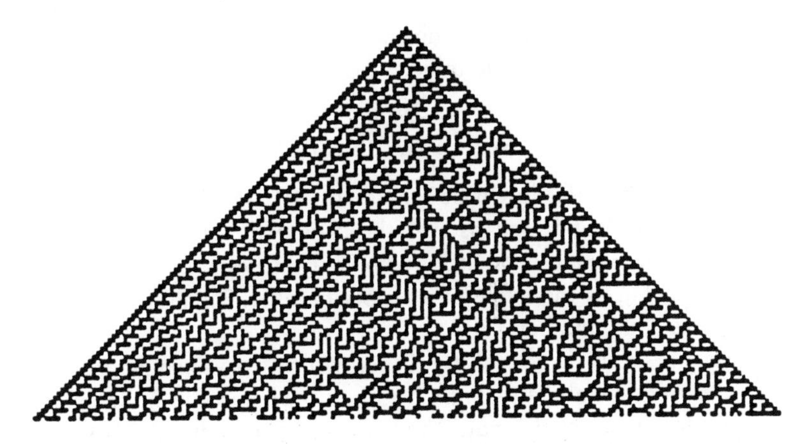

Fig. 140 A simple CA can evolve a deep pattern from a simple initial state. From Stephen Wolfram, "Origins of Randomness in Physical Systems," *Theory and Applications of Cellular Automata,* World Publishing Company, Singapore, 1986.

head would currently be if it were moving as before. At the end of each time tick, the "on" triangle would mark or erase its cell, set one of its neighbors to a selected nonzero state, and then turn itself off. Looked at in this way, a K-state Turing machine is a $2K$-state 1-D CA.

Recall that John Horton Conway helped prove that the game of Life can be thought of as a universal Turing machine, even though it is only a two-state CA. Wolfram is currently trying to prove that there is a one-dimensional analog of Life. That is, he is trying to find a two-state, 1-D CA rule that is a universal Turing machine.

The two-state, 1-D CA rules are indeed of surprising power. To demonstrate this, I'd now like to show Wolfram's catalog of the thirty-two "legal and totalistic" rules that determine a cell's next state on the basis of the present state of the cell's "two-neighborhood," which consists of the cell and its two nearest neighbors to either side — five cells in all. A rule's being "legal" means that, if a cell's two-neighborhood is all blank, then it stays blank. A rule's being "totalistic" means that a cell's next state is determined solely by the sum of the number of marks in the cell's neighborhood. This sum can range from zero (no cells marked) to five (all cells marked).

A totalistic, two-state, two-neighborhood, 1-D CA rule like this can be specified by a simple table:

Sum	5	4	3	2	1	0
Next state	0	1	0	1	0	0

The code of the rule shown here is, then, "010100." The condition that a rule be "legal" means that the code ends with a 0. We can describe such a code even more compactly if we regard the zeros-and-ones string as the name of a binary number. Thus, the rule just shown has code number "20."

There are thirty-two legal, totalistic, two-state two-neighborhood, 1-D CA rules in all; and Fig. 139 shows what some of them do to a randomly selected input. The way to interpret these pictures is to regard the top line as the initial one-dimensional CA pattern, the next line as the pattern after one step, the next line as the pattern after two steps, and so on. The rightmost cell is thought of as connecting to the leftmost cell. The algorithmic complexity of these patterns is low, but if we use a long tape and start with only a few cells marked, then the logical depth of the line pattern after *N* units of time will typically be about *N*.

We have already mentioned the difficulties in predicting what a

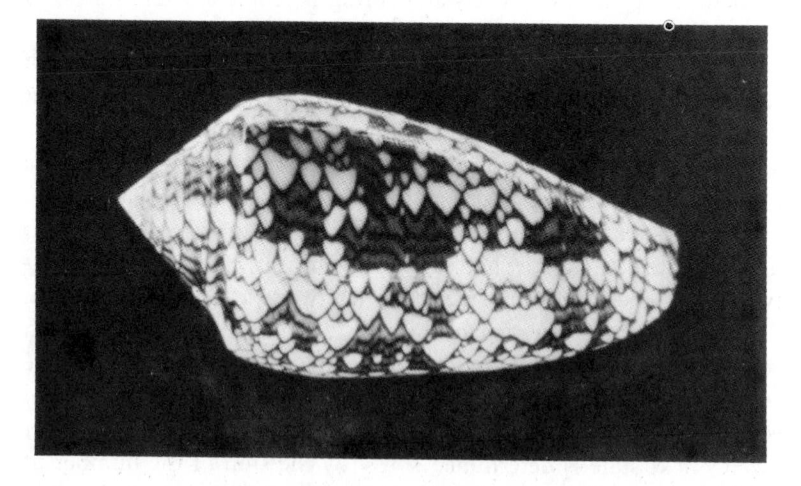

Fig. 141 A cone shell. From Stephen Wolfram, "Cellular Automata," *Los Alamos Science*, Fall 1983.

universal computer will do with a given input. The halting problem, in particular, says that many times there is no better way of predicting a computation than to actually run it, or to simulate it in some way. Wolfram speaks of such computations as *irreducible*. Many CA's seem to carry out irreducible computations — meaning that there is no quick way to predict what the pattern will look like after so and so many steps. Wolfram feels that many physical processes are, in effect, irreducible computations. This means that we can have simple starting conditions whose consequences we are unable to predict.

One specific class of physical examples of CA-like patterns can be found on certain seashells. The cells along the edge of the these molluscs' shell-growing tissues have a kind of CA rule for determining what color the next bit of shell should be. Thus the presence of one black-producing cell among a cell's immediate neighbors might cause the cell to produce black, the presence of two black-producing cells could inhibit the same response. Where D'Arcy Thompson found a mathematical explanation for shells' shapes, Stephen Wolfram seems to have found a mathematical explanation for some shells' coloring. It is likely that similar processes are involved in the growth of plants.

Just as we can talk about one-dimensional cellular automata, we can talk about CA's of 3, 4, or even ω dimensions. The great vision among present-day information theorists is that it may be reasonable to regard the world as a huge CA carrying out irreducible computations.

In such a cellular view of the world it seems appropriate to take a fractal kind of approach and not demand that there be one best minimal cell size. In a society a "cell" is a person. In an organism a "cell" is a cell. In a very small region of matter a "cell" is an atom. Whether or not there is some ultimate minimal-size cell is perhaps not so important. It may be better to think of the world as being structured on a hierarchy of levels. The higher levels depend on the lower levels, but the rules of operation of the levels close to us in size are easier to understand.

Present-day physicists feel themselves to be close to some master equations that will describe all physics. Does this mean the end of science? Of course not. If we think in terms of CA's we can see what's going on a lot more clearly.

Finding the "master equation" is like learning the simple rule that governs the behavior of each individual cell in the CA. How can simple rules lead to complex phenomena? Via long runtime. If the computation is irreducible, then there is no way to predict the phenomena that arise.

This, then, is the old solution to the split between our search for simple rules and the intricacy of the world we actually live in. It is in fact possible that the world has evolved like a $U(P) = M$ computation from some very simple initial state. The world may indeed have a low information-theoretic "complexity." But the world has a very high "logical depth." There is no quick and dirty way to predict things about it. Simplicity of conception and intricacy of execution are not antithetical.

People sometimes say, "If the world really has a simple explanation, then where did all the randomness and information come from?" We can think of the information as coming in with the passage of time. Each tick of time increases the depth of the world computer's output.

My purpose in writing *Mind Tools* has been to see what follows if one believes that everything is information. I have reached the following (debatable) conclusions.

1) The world can be resolved into digital bits, with each bit made of smaller bits.
2) These bits form a fractal pattern in fact-space.
3) The pattern behaves like a cellular automaton.
4) The pattern is inconceivably large in size and in dimensions.
5) Although the world started very simply, its computation is irreducibly complex.

So what is reality, one more time? An incompressible computation by a fractal CA of inconceivable dimensions. And where is this huge computation taking place? Everywhere; it's what we're made of.

INDEX